Law and the Mind

Volume 184 Sage Library of Social Research

RECENT VOLUMES IN . . .
SAGE LIBRARY OF SOCIAL RESEARCH

13 Gelles **The Violent Home, Updated Edition**
78 Matthews **The Social World of Old Women**
84 Gelles **Family Violence, 2nd Edition**
94 Rutman **Planning Useful Evaluations**
126 McPhail **Electronic Colonialism, 2nd Edition**
142 Toch/Grant **Reforming Human Services**
150 Frey **Survey Research by Telephone, 2nd Edition**
152 Estes/Newcomer/and Assoc. **Fiscal Austerity and Aging**
153 Leary **Understanding Social Anxiety**
154 Hallman **Neighborhoods**
155 Russell **Sexual Exploitation**
156 Catanese **The Politics of Planning and Development**
157 Harrington/Newcomer/Estes/and Assoc. **Long Term Care of the Elderly**
158 Altheide **Media Power**
159 Douglas **Creative Interviewing**
160 Rubin **Behind the Black Robes**
161 Matthews **Friendships Through the Life Course**
162 Gottdiener **The Decline of Urban Politics**
163 Markides/Mindel **Aging and Ethnicity**
164 Zisk **Money, Media, and the Grass Roots**
165 Arterton **Teledemocracy**
166 Steinmetz **Duty Bound: Elder Abuse and Family Care**
167 Teune **Growth**
168 Blakely **Planning Local Economic Development**
169 Matthews **Strategic Intervention in Organizations**
170 Scanzoni **The Sexual Bond**
171 Prus **Pursuing Customers**
172 Prus **Making Sales**
173 Mayer **Redefining Comparative Politics**
174 Vannoy-Hiller/Philliber **Equal Partners**
175 Brewer/Hunter **Multimethod Research**
176 Chafetz **Gender Equity**
177 Peterson **Political Behavior**
178 So **Social Change and Development**
179 Gomes-Schwartz/Horowitz/Cardarelli **Child Sexual Abuse**
180 Evan **Social Structure and Law**
181 Turner/Turner **The Impossible Science**
182 McCollum **The Trauma of Moving**
183 Cohen/Adoni/Bantz **Social Conflict and Television News**
184 Gruter **Law and the Mind**
185 Koss/Harvey **The Rape Victim**
186 Cicirelli **Family Caregiving**

Law and the Mind

Biological Origins of Human Behavior

Margaret Gruter

Sage Library of Social Research 184

SAGE PUBLICATIONS
The International Professional Publishers
Newbury Park London New Delhi

To my children
And theirs

For information address:

SAGE Publications, Inc.
2455 Teller Road
Newbury Park, California 91320

SAGE Publications Ltd.
6 Bonhill Street
London EC2A 4PU
United Kingdom

SAGE Publications India Pvt. Ltd.
M-32 Market
Greater Kailash I
New Delhi 110 048 India

Printed in the United States of America

Library of Congress Cataloging-in-Publication Data

Gruter, Margaret.
Law and the mind : biological origins of human benavior / Margaret Gruter.
p. cm. — (Sage library of social research; v. 184)
Includes bibliographical references.
ISBN 0-8039-4045-9. — ISBN 0-8039-4046-7 (pbk.)
1. Law and Biology. I. Title. II. Series.
K328.G78 1991
340'.11—dc20 90-23412
CIP

FIRST PRINTING, 1991

Sage Production Editor: Astrid Virding

Contents

The legislator "should depart in no way from what is most noble and most true; but when some aspect of things turns out to be impossible . . . he should steer away and not do it. Instead, he should contrive to bring about whatever is the closest to this from among the things that remain."

Plato, *Laws,*
V 746 b-c:trans. T. L. Pangle
(New York: Basic Books, 1980), p. 134.

Prologue

In the last 30 years there has been a rebirth of interest in approaches to law that may appropriately be called "Law and . . . ." These approaches all look to other disciplines for the insights that they can give to law. At their most sophisticated they also use law as a way of furthering, of asking deeper questions about, the related disciplines. Although the most dramatic and widespread use of such an approach has been seen in the growth of the New Economics Analysis of Law, Law and Philosophy, Law and History, Law and Literature have all also been the subject of major recent renewals. Other areas like Law and Psychoanalysis have remained quite important even if no major new movement has centered around them in the past few years.

Oddly, however, until recently relatively little has happened in deepening the relationship between law and some fields that should be of special significance to law. I am thinking of evolutionary theory, of biology generally, and of those historical-cultural studies that can, albeit with some discomfort, be gathered under the heading of anthropology. I have found this gap not only strange, but quite disturbing. In my own work, I have often felt that the explanations that have been given (by me as well as others) for certain legal rules, or lines of authority, in terms, say, of economics seemed remarkably frail, brittle, and bloodless. This has caused me to ask whether a fuller, richer explanation could not come about if we were, for example, to incorporate anthropological insights into the analysis. Indeed, my sense of the importance of these approaches has led me to suggest to young people starting out to be legal scholars, that, were I their age, I would learn about these disciplines in order to discern what they could teach us about law (and what law would require, and teach us, about them) rather than, as I did some 40 years ago, study economics.

It is for these reasons that Doctor Margaret Gruter's book is one that I particularly appreciate. My distinguished colleague, E. Donald Elliott, will discuss her approach and work in detail in the Foreword, and so I will restrain myself from commenting on the fascinating insights she presents. While I, like Don, do not agree with everything she says, I applaud the enterprise on which she has embarked. I cannot, however, keep from mentioning her outstanding support, through the Gruter Institute for Law and Behavioral Research, for widely varying studies that in different ways examine the links between law, evolution, and ethology. When a person has given so much of her time and energy to furthering the work of others, it is always a special pleasure to find that it has not been at the cost of her own scholarship. For this reason, too, the publication of this book is a most joyous event, and it is with unusual delight that I welcome it.

—Guido Calabresi
Dean, and
Sterling Professor of Law
Yale Law School

Foreword

This is a remarkable book by a remarkable woman. For almost two decades, Margaret Gruter has been the chief prophet, impresario, patron, and guiding spirit behind a pioneering effort to bring the insights of modern behavioral science, particularly biology, into the law. The focal point for her work to marry law and biology, both in Europe and in the United States, has been the Gruter Institute for Law and Behavioral Research, which she founded in 1982 (and on whose advisory board I now serve).

The mission of the Gruter Institute has been to stimulate scholarly collaboration among researchers in law, the sciences, and related disciplines such as political science and economics. The results, both scholarly and personal, have been impressive. Over 100 scholars have attended conferences or other events sponsored by the Gruter Institute, a small enterprise that Margaret and one assistant run out of her home, and the list of publications growing out of these efforts is now more than a dozen books and growing rapidly.

This book is Dr. Gruter's personal contribution to the developing field of *law and biology*, which she, more than anyone else, helped to create. In it, Margaret outlines her own personal vision of what is particularly salient in modern biological theory for the law, and then attempts to apply these findings to two specific areas of the law, family law and environmental law.

The first three chapters summarize some of the key findings of modern behavioral sciences, particularly evolutionary biology and ethology (the science of animal behavior, derived from the work of Konrad Lorenz). This is not the simplistic evolutionary biology, bloody "red in tooth and claw," that we learned in high school. Gone are simplistic slogans such as "survival of the fittest." Rather, a crucial theme in modern evolutionary biology is altruism and cooperation—how these helping behaviors develop, and how they enable humans to live together in large, complex social groups.

Thus through the work of biologists such as Robert Trivers and Richard Alexander, anthropologists such as Lionel Tiger and Robin Fox, and political theorist Roger Masters, ethology and evolutionary biology are beginning to propose *empirical* answers to questions about the sources and nature of human values. These developments in evolutionary theory hold out the prospect of a convergence with parallel insights in coordination theory that are developing in economics, game theory, and jurisprudence in recent years. The resulting synthesis, the "new naturalism," addresses the nature and functions of norms in a way that holds out the prospect of making real progress on issues that when addressed *a priori* rather than empirically have bedeviled moral philosophy for 2,000 years.

Of course, a crucial question, perhaps THE Question, that all work in the law and biology must confront is: "What relevance do studies of *animal* behavior have for *human* society?" Dr. Gruter's approach to this issue must not be confused with vulgar sociobiology. She does not assert that all human behaviors are determined by biology, allowing ample room instead for cultural and other factors. Moreover, Margaret is also well aware of the great diversity of humankind, and acknowledges at several points in these pages that variations in characteristics among individuals within populations may be as significant for particular problems as the commonalities built into them by their common evolutionary history. By avoiding the errors of oversimplification that infect much of the work by lawyers who dabble in biology, Margaret demonstrates her competence in understanding biology.

She also demonstrates her competence as a legal scholar by avoiding the "naturalistic fallacy." Unlike some law school professors who write about sociobiology and the law, Dr. Gruter does not assume that those behaviors that have a basis in evolutionary biology ought necessarily to receive some special normative or privileged position in the law. On the contrary, hers is the milder, more sensible position that human beings are predisposed—"programmed" is her favored word—to certain behaviors by their evolutionary heritage. Law can go against the evolutionary grain, but only at some cost. Lawmakers should, therefore, she contends, at least understand what behavioral predispositions are built into our species by our evolutionary past, even though we may choose to use law to try to counteract these natural tendencies.

Indeed, the essence of Margaret's vision is of the divided self. She sees two opposing spirits built into the human soul by our evolutionary past, one selfish and violent, the other altruistic and loving. Law, she believes, is one of the devices that human beings have developed to mediate between these two sides of the human soul.

In the final two chapters, Dr. Gruter illustrates her approach by applying it to two specific bodies of law, family law and environmental law. Specialists in these two fields will undoubtedly find much with which to disagree in her analysis. Those wise enough to read for what they can learn will also discern many provocative, original ideas. For example, in my own field of environmental law, Dr. Gruter contends that the central problem that defines the field is that the benefits of investments in environmental protection cannot be seen within a human lifetime.

Some might quibble that, as a matter of fact, in some areas of environmental management, discernible progress has been made in just the last 20 years; once-polluted rivers like the Potomac now again support fish and are safe for recreation;

the air in many of our major cities is measurably better, and so on. Others might attack from a different direction, pointing out that many other areas of law, such as property, or trusts and estates, not to mention constitutional law, also deal with time periods longer than the human lifespan.

There is some force to these points, and many more like them, that go to the details of Margaret's argument. Some might wish that she had had the time to anticipate every possible objection and take them all into account one by one, in the careful, methodical manner characteristic of the law reviews. But that carping, narrow style of scholarship would have been inconsistent with Margaret Gruter's mission, which is to lay before us a vision of how law and biology may come together.

The great value of this book, like the value of much of her life's work, lies not so much in the specifics as in the vision itself. Dr. Gruter's insight that the core problem of responsibilities to future generations defines the field of environmental law is basically right. By approaching this issue (which is not unfamiliar to those working in the field) through the lens provided by the biology of altruism, she brings a new and distinctive focus to the field. It remains for other scholars to pursue her lead to explain how and why law can address intergenerational problems successfully.

Dr. Gruter's book is a pioneering work. Like all pioneering efforts, she does not complete a field. She begins it.

—E. Donald Elliott
Professor of Law,
Yale Law School

Preface

This book is written to introduce legal scholars and practitioners to the importance of biology and its relevance to a wide range of legal phenomena.

What is the connection between biology and law? One answer is that law is one of the creations of the human brain or mind (the terms "mind" and "brain" will be used interchangeably in this book). While serving the needs of our ancestors for several millions of years, the brain has evolved as a composite—a triune brain as Paul MacLean calls it—made up of a mixture of ancient and modern parts. It has been the originator of rules and has shaped behavior throughout human history. Our existence testifies to the fact that the mind has mastered the many tasks associated with developing, changing, and adapting rules, creating legal systems that could and can accommodate the challenge of ever-fluctuating environments. Most important for the purpose of this book, the human mind has been able to maintain a balance between change and continuity, a balance that is essential for human survival and successful competition with other species.

From the perspective of evolutionary theory, law is a creation of many minds, past and present. Because of this relationship, law is unavoidably linked to our biological nature. This book focuses primarily on one area in contemporary biology—the study of animal behavior (or, as it is generally called, ethology). Many other subfields in biology also have important implications for our legal system. In the past generation, for example, we have seen revolutionary changes in our understanding of behavioral ecology, ethology, genetics, gene therapy, neurophysiology and neurochemistry, population genetics, and ontogenetic processes. No one person can claim to speak knowledgeably in all of these disciplines. Ethology—the subject of my own work over the last 20 years—is particularly relevant to law, however, because of its focus on the behavior of both human and nonhuman species. By studying the way animals like ourselves behave when living in groups (which is the central concern of the law), we open the door to insights into human law.

As it is used here, the word *law* references the legal system, specific laws and their enforcement, and above all, legal behavior. Over time, law has acquired a "life of its own," and like a living organism, law grows in response to ideological and religious beliefs as well as rational considerations originating in different parts of the brain (e.g., limbic system and neocortex).

Being aware of these biological connections is necessary for understanding the view of the legal system that this book develops. Knowledge is neutral. What we do with knowledge is what matters. Nowhere is this point more relevant than in interactions between biology and law. Our understanding and application of the law can be enhanced significantly by assimilating findings from biology and related sciences. Yet the stream of knowledge runs both ways, for law also alerts biologists to certain issues that need to be addressed because of their social and legal implications. This interplay deserves our close attention and appreciation. Law and biology advance at their own pace, however. The pace of biology now exceeds that of law. We live in an era in which findings from biology raise fundamental questions for the law. How knowledge from biology is accepted and incorporated into law becomes an issue of immense social importance. Scientific research, with the assistance of law, holds great promise for improving our lives. Scientific research, without the guidance of law, is dangerous.

A basic proposition of this book is that legal research and practice can keep pace more effectively with changes in human society when findings from the biological sciences are known, understood, and incorporated into legal thinking and practice. To make this proposition a reality, our "brain child"—the law—must be reconnected with its origins in human behavior. Biology provides us with both an orientation and the insights with which to carry out this task, namely, to reevaluate rules and laws, their functions in society, and their effectiveness.

The genesis of this book goes back to the 1940s. It is the product of a personal search to bridge the gap between biology and law, to reconcile the human yearning for justice and a just society with the realities of this century as I have experienced them.

I chose law as my calling, motivated by a desire to learn about justice. My education as a law student began at the University of Heidelberg, Germany, shortly before the outbreak of World War II. Those were unusually difficult times for a young woman who was interested in philosophical concepts of fairness, justice, and ethics. Confronted with the cruel, mind-boggling reality of a dictator determined to rule the world, academic endeavors were often a secondary priority.

My way of thinking was shaped by two experiences. The first was my education in the Humanistische Gymnasium prior to matriculating to the Ruperto Carola University of Heidelberg. The second was my father, who wanted to give his only child the advantages of a liberal education. The Gymnasium, attended mostly by boys, was nine years of intense schooling. During those nine years, many students left, some continuing their studies in schools with less demanding curricula. Of the more than 60 students who, in 1928, had entered the Gymnasium at the age of 10, only 14 (11 boys and 3 girls) graduated in 1937. There were no optional subjects. We were saturated with the work of Greek and Latin authors, all read in their original languages, and the history of the Ancient World—the philosophies and the ethics on which Western civilization has been built. Passing grades were required in all subjects.

The changes from 1933 to 1937 affected the curriculum in some subjects, yet the general liberal trend of the Greco-Roman tradition of our education continued. After 1933, there was greater emphasis on athletic achievements and sports, and the teaching of modern history was slanted toward the propaganda of the new regime. Books by Jewish authors were banned from classes in German literature. The teachers (with the exception of some who were generally disliked because they had gained their positions due to party membership) tried to evade excesses of party-line teaching. They were only partially successful. The reality of a dictator ruling the nation of "Dichter und Denker" left none of us untouched. We experienced the dark side of life in Germany during the 1930s, with its militaristic spirit and its cruelty. We had to march in step whenever the Party called.

A few inspiring teachers in the Gymnasium, later at Heidelberg University, and, later still, at Stanford University, encouraged my curiosity and desire to question both ideological and religious value judgments and views of reality. Through it all, there remained a deep commitment to pursuing the elusive ideal of justice.

A series of personal experiences led me to return to the university 25 years after writing my dissertation. It is not surprising that I brought to this effort many questions that had remained unanswered for the preceding two and a half decades. In particular, I had not resolved the contradictions of being a liberal arts student who, on one hand, was steeped in the ethics and philosophies of a classical education and, on the other, had lived with the realities of Germany during the 1930s.

I have been married to a physician since 1940. The oldest of our children, also a physician, had been involved in biological research since the time she entered college. Following our immigration to the United States in 1951, we lived for 18 years in an intellectually isolated section of rural Ohio. I led the life of a country doctor's wife, assisting my husband in all aspects of his practice.

Motivated partly by sheer boredom, I assumed responsibility for the development and administration of a medical facility that, by 1970, housed 150 mentally retarded persons of all ages. During the Ohio years, my most exciting intellectual moments were provided by my husband and daughter in our discussions of biology. These discussions were often unpleasant and disturbing for they threatened many of my values, particularly those dealing with the power of the human mind and human abilities. I was surrounded daily by mentally retarded individuals. This environment attested to the truth of many of the biological findings in the scientific literature that regularly found its way to the desks in our home. Living in this environment convinced me that psychological explanations were insufficient either to understand or to treat effectively the mental disabilities we encountered. It became clear that such explanations, as well as social therapies intended to improve the abilities of patients through environmental alterations, produced very limited results, even with the help of the then available medications. Many patients could not be trained in the simplest tasks of bodily hygiene or self-nourishment. A few could learn simple tasks and develop a limited vocabulary. Fewer still learned to perform satisfactorily household and gardening activities. The patients received all the nurture imaginable, yet the limits of their improvements were largely set by their brains. It was in this setting that I more fully grasped the close interrelationships between brain and behavior. I could no longer argue that a scientific approach to human behavior—including legal behavior—was possible without a basic understanding of biology.

In 1969 we left Ohio and moved to California. I was soon attracted to the possibility of again becoming a student. After auditing courses at Stanford University Law School, I was admitted in the Master of the Science of Law (J.S.M.) program. There I could create a menu to satisfy my intellectual appetite.

My interest in the writings of Konrad Lorenz and other ethologists soon led to contacts with members of the Department of Psychiatry and Behavioral Sciences. In 1969, David Hamburg was department chair and Jane Goodall was guest professor. Her studies of chimpanzees were instrumental in my turning to look at other species. In 1972, I visited with Konrad Lorenz at Seewiesen. He encouraged me "to build the bridge between ethology and the law." Subsequently, he often sustained me in my efforts to navigate the many fields dealing with law and human behavior toward a concept of the ethology of law.

In retrospect, a somewhat naive approach to the study of law may have been a necessary condition for me to proceed. I could appreciate the law both as a creation of the human mind and as a product of the biological mechanisms that support and make possible the human quest for order and justice. This dualistic view made perfect sense to me, although I soon learned that there were detractors. It seemed so clear to me that an understanding of evolutionary biology was a prerequisite for understanding our rules and laws. It seemed equally clear that early human cultural achievements of our ancestors pointed to the important role of law in the shaping of *Homo sapiens*. Also apparent was the beauty of law as a structure, comparable to a well-designed architectural edifice, or to a work of art or nature that evokes a sense of awe. I labored to understand these phenomena. One answer to these efforts came when I recognized the pleasing feelings associated with balance, with continuity and unity, and their relationships to processes in the human brain that induce satisfying central nervous system states through our senses. Humans, the most versatile and adapted of the primates, come together on a common ground in their desire for order and justice. Findings from studies of child development, the origins of language, and the acquisition of language in the life of individuals, served to further my understanding of the human need to find rules in order to cope with the diverse stimuli of our world. Rule-making, rule-incorporating, and rule-obeying are integral parts of each individual's development. Clearly a vast number of biological mechanisms are involved in the events of such development, a subject to which I will return presently.

The view that a behavior has biological origins does not imply that human behavior is ontogenetically fixed, that learning does not occur, or that deterministic theories of human behavior are valid. Increasingly, research points to a surprising degree of behavioral plasticity among nearly all species. Yet, research also points to the fact that many behaviors are biased by our genetic makeup and that there are constraints on plasticity. These facts underscore the importance of addressing biological findings as we attempt to understand the realities within which legal systems must operate and what they might obtain.

In 1972, I looked up my almost-forgotten dissertation of 1944 on English Divorce Law. I was surprised and delighted to find that even then I had come to the conclusion that

> regardless of any changes made in the divorce law meant to equalize the chances for men and women, the law cannot create complete equality, if only because of biological differences between men and women. Attitudes and responsibilities towards their offspring differ across sexes and these differences appear to be based in important ways on their biological differences. The consequences of even the best intended equal treatment under the law will be different for each sex.

What I have observed since then has confirmed this view. Yet one must be cautious. Exploring the possible legal implications of biological differences between the sexes does not a priori imply that the law should give preferential treatment to one or the other sex. Nor does it imply that women and men should cease in their efforts to obtain equality and justice. What exploring biological differences does mean is that many parts of law (e.g., family law, environmental law, contract law) may be informed by findings from biology and that efforts to arrive at solutions that are workable and contribute to both individual and social welfare are more likely to be successful if these findings are included in discussions of planned reforms.

Am I still searching for a satisfactory definition of justice? Yes! What has this 20-year search taught me? I believe that available evidence and ideas are sufficient to make a compelling case for a biological explanation of justice. The path to this conclusion is not without its detours, however. It is important, for example, to distinguish between the concept of justice held in a given society and one's personal sense of justice. The former is a product of interactions between many individual responses to questions of fairness and the social reality within which these questions are asked. The latter is the result of one's personal history, one's culture, environmental variables, and brain chemistry. What one society calls just and fair does not always apply universally. Yet, underlying these differences are common biological properties, and on important issues there often are cross-societal agreements on what is fair and what is not.

At present, research in neuroanatomy, neurotransmitters, and other brain research is providing us with information on the anatomical network, the chemistry, and the mechanisms that are instrumental in individual value judgments. In the next decades such research is likely to provide information that will help us understand how, under differing environmental conditions, a given individual society will choose the concept of justice by which to live.

What we are discovering at this time, and where present research is likely to be helpful, is that there are brain mechanisms that can in large part explain why individuals develop a sense of justice. Currently, there is evidence to support the view that specific brain mechanisms direct individuals toward achieving balance and structure in their lives. Legal behavior mirrors these efforts. The same point holds for the satisfaction individuals derive from remaining within a given social framework. Type of social membership influences legal behavior. Assuming that the mechanisms that shape a sense of justice in the individual brain are present among all humans, which variant of the concept of justice a person embraces will depend on prior experience, especially during preadult life.

I began this book in the early 1970s. At that time, I searched the libraries for findings to support or refute my ideas. The classical writers in law obviously did not have access to the information then emerging in ethology and brain research, the primary areas on which my conclusions are based. Eugen Ehrlich's "living law," Malinowski's anthropological definition of legal rules, Adamson Hoebel's descriptions of law as behavior, and Paul Bohannan's work on concepts of justice among the Tiv, helped to bridge the gap that separates law and the natural sciences from the law side. Among the early writers of the twentieth century (as well as among my contemporaries who published in legal philosophy, sociology of law, anthropology of law, and related fields), there was scarcely a reference to biological approaches to law and literally none about neurological findings dealing with human behavior as it relates to the law. Aristotle had based his thoughts about man as a "zoon politicon" on the insights of the natural sciences more than 2,000 years earlier. His intuition and conclusions often seemed to be more relevant than much of the literature written since then. There were some exceptions, of course. The *Ethical Animal* (Waddington), *The Imperial Animal* (Tiger and Fox), and similar rare finds were oases in an otherwise barren desert. Each of these authors recognized humanity's biological nature and discussed the implications of this nature for social organization.

The basic ideas that guide this book were published in a German version in 1976. In retrospect, the 1976 publication appears to be the first book seriously to embrace possible implications of modern biology for law. In this edition, I have added new materials. Then and now, the book was not and is not intended to be an exhaustive treatment of either biology or law. It expresses my view of general trends in biology and law following a decade of studies and thought in the field. The book can best be understood as an attempt to set the direction and tone of future research.

Several friends have spent time reading through early drafts of this book. Their reactions made it clear to me how controversial this book will be, even among scientists. I have tried to focus on facts that can be supported. Yet, I have a point of view, a personal opinion on the very intimate issues discussed in this book. In part it is the view of a woman, a mother, a wife, and a grandmother. But it is also in part the view of a person who has spent time exploring scientific evidence and relating it to the prevailing law. And it is also in part the view of a person who has given much thought to those problems that confronted her as a doctor's wife in the postwar era of Heidelberg and in rural Ohio in the 1950s and 1960s. When I was working with my husband, the women patients who came for help came from all walks of life. In Heidelberg, they ranged from the prostitutes whose business was flourishing during the postwar American occupation to very ordinary middle-class married women, usually well educated, mothers of several children, yet sometimes unwilling or unable to confront another pregnancy. In Ohio, they were mainly the wives of farmers and blue-collar workers. These experiences, combined with the work with the retarded persons, have shaped my view.

Those scholars and scientists (Lawrence Friedman, Jane Goodall, David Hamburg, Manfred Rehbinder, Michael Wald, and Wolfgang Wickler) who read the 1976 German book in draft form were my associates during the 1970s, and I am grateful for their help in those early years. It would not have been possible to sort through the tremendous amount of new research that became available during the 1980s and to rewrite the book accordingly without the support of additional "kindred spirits". These are my associates at the Gruter Institute who provide intellectual stimulation and help in lifting the veil of ignorance we humans share. The participants in the Gruter Institute conferences all are part of this group and I am grateful for their contributions, which have shed light on various subjects.

The new drafts of the book were read by E. Donald Elliott, Roger Masters, and William Rodgers, who gave constructive criticism and suggestions. Paul Bohannan, my coeditor in *Law, Biology and Culture* (1983), spent much time guiding me through one of the many early drafts of this book. For help with the final version of this book I am greatly indebted to Michael

McGuire, who answered many questions pertaining to biology and encouraged me in the seemingly endless struggle to put it all together.

Finally, the book would not ever have been started without my husband, who opened my mind to the natural sciences, and without my daughter's encouragement to pursue this course of study. Last but not least, my thanks go to Gerti Dieker, my assistant for more than two decades, who again has shown endless patience and great skill in getting the book ready for publication.

PART I

Historical and Conceptual Background

ONE

The Evolution of Law in Ethological Perspective

The origins of law,[1] like the origins of social order, are rooted in the behavior of early humans and their primate ancestors. It is these relationships and their implications for law that are the central topics of this book. This chapter sets the framework for this inquiry. I will begin with a discussion of man-made law.

MAN-MADE LAW AND LEGAL BEHAVIOR

Man-made laws are ideas and commands expressed in words. They are developed by the mind.[2] Mind and law interpret the past, the present, the future, and they anticipate both known actions and as yet unexperienced events. Man-made law is unique in terms of its complexity, the broadness of its applications, its influence on organizational structures and on the lives of individuals. Man-made law can and does adapt to the necessities of a changing world, although it often lags behind rapid changes. At any given time, the law can and will decree changes that are deemed necessary or opportune by those who make laws. Yet the effectiveness of all laws, old or new, depends on the ability and willingness of large groups of individuals to alter and adapt their behavior to laws often devised by a small rule-making minority. The effectiveness of changing or regulating society by law depends on legal behavior. Legal behavior is one of many behavior traits that have evolved among humans. Individuals can obey the law, evade or ignore it, or disobey it. Compliance and deviance are both legal behaviors, representing an interaction of the individual with the legal system. A law is effective if the majority of those to whom the law is addressed obey the law. The larger the majority, the more effective is the law.

Legal behavior did not evolve in a vacuum. It evolved in conjunction with other behavioral traits. For tens of thousands of years—the period of recent human evolution—humans lived in groups ranging from 50 to 200 people.[3] To survive, they had to find ways to raise their offspring until they reached social and biological maturity. Mothers had to reassure, direct, teach, and adjust to their infants. Complex language had to develop. Information about the environment, known at first to only a few, had to be communicated—and rules for sharing had to be developed and modified.

Early in human evolution, rules that infants had to obey (if only because they could be enforced by their mothers or other group members) were critical for survival. Rule-following thus was intertwined with other behaviors essential for survival. Together, they proved adaptive. Rule-making and rule-following behaviors date back millions of years among hominoids and perhaps as much as 100,000 years among *Homo sapiens*.[4] During the latter period, dispositions for rule-making and rule-following behavior were likely to be favored by selection, with predispositions for these behaviors becoming an integral part of the species genome. Fox[5] puts the matter this way: our ancestors were "selected for speaking, classifying and rule-making creatures who could apply these talents directly to the breeding system. Whatever it was that was succeeding became built into the cortical processes. This included the intellectual and emotional apparatus." Tiger[6] comes to similar conclusions in discussing interactions between kinship, food sharing, and early ethics: "wired-in emotional components of mammalian feeding are extended and generalized to the wider social network." No doubt there were numerous other factors that contributed to our capacity to seek and, in turn, to follow the rules we created. Once developed, however, they set the direction of subsequent human history.

CONCEPTS OF LAW AND SOCIAL ORDER

Concepts of law and social order (norms) are often confused. They are distinct, however. Moreover, the details of their distinctiveness are of more than passing interest in the study of the origins of law. Bronislaw Malinowski[7] supplies a broad definition of law in the following:

> The rules of law stand out from the rest in that they are felt and regarded as the obligations of one person and the rightful claims of another. They are sanctioned not by a mere psychological motive, but by a definitive social machinery of binding force, based, as we know, upon mutual dependence, and realized in the equivalent arrangement of reciprocal services, as well as in the combination of such claims into strands of multiple relationships. The ceremonial manner in which most transactions are carried out, which entails public control and criticism, adds still more to their binding force.

In this definition, an important factor in distinguishing laws from other rules or norms is the existence of a social machinery that serves as a binding force.

Although other species follow rules—nonhuman primates, for example, contest dominance status in predictable ways and seldom injure each other during the process—legal systems directing individual and group behavior are the results of interactions between those who make and administer laws and those to whom the law is addressed. Some find it advantageous to obey laws, or at least do not openly break them. Others may respond in different ways. The mere number of professionals involved in one or another aspect of the legal system (which has been and is increasing rapidly, especially in Western societies) attests to both the complexity and the importance of legal behavior. Humans seek, develop, and elaborate rules. They evoke the symbol of justice. They preach the dogma of equality before the law. And, in most instances, they obey the law. Yet, to be effective, law cannot stray too far from behavior. Rules that are at cross-purposes with strong biological predispositions—such rules are often imposed for ideological or religious reasons—are either short-lived, disregarded, or both. The fortunes of the Volstead Act illustrate this point. The Volstead Act, or the National Prohibition Act, was approved by Congress in 1917 as the Eighteenth Amendment to the U.S. Constitution. The act was formally passed by Congress in 1919, and enforcement took effect on January 16, 1920. The amendment was repealed in 1933. The act turned out to be a case study in misguided legal efforts to force behavioral change. It failed to accomplish its desired result, that of eradicating the social consequences of excessive alcohol consumption. Despite vigorous efforts at enforcement, thousands of citizens at all levels of society colluded with racketeers to acquire alcohol. Not only did this attempt at changing human behavior turn out to be largely unenforceable, but it also created unforeseen consequences in that attempts to prohibit alcohol consumption encouraged many citizens who normally obeyed laws to work outside the law.[8]

NATURAL LAW

For at least 3,000 years, philosophers and jurists have endeavored to fashion legal systems that facilitate a reasonably peaceful and predictable social order. Man-made laws have been one of the chief tools in this endeavor. Everyday experience and philosophical, religious, and ideological dogmas, along with insights developed from the natural sciences, have contributed to these developments.

A number of theories dealing with the nature and the origins of law characterize different historical periods. These theories continue to influence contemporary legal theory, lawmaking, and the administration of law. Even the briefest account of these theories is beyond the scope of this book. However, it will be useful to examine selected examples from history because of the perspective they provide on our current attitudes and assumptions about law. Those to be discussed deal with the role of law and the concepts of nature and evolution. In this review, it should be noted that when the authors use the terms *nature* or *natural law* and draw conclusions from their ideas about evolution and nature, they have often been unfamiliar with the findings of biology.

In pre-Socratic Greece, early cosmologists such as Thales and Antiphon defined and emphasized the difference between nature (physics) and law or convention (nomos). Whereas earlier religious or social traditions had not recognized the fundamental status of this distinction, these Greek thinkers insisted that the varying customs and laws of each society are basically different from the rules of nature, which are everywhere the same.[9] As Aristotle put it with characteristic simplicity in *Ethics*, customs such as the cost of ransoming a prisoner differ from place to place although "fire burns both here and in Persia."

The Sophists (represented by Thrasymachus in Plato's *Republic*) reasoned that human laws were merely conventions or agreements based on self-interest. There thus was no natural basis for the laws and customs adopted in any one society (beyond the self-interest of the rulers and the necessity to survive of the powerless). Plato, and his student Aristotle, countered by asserting that there is a substantive definition of right or justice that is "according to nature."[10] This tradition gave rise to the concept of natural law or *jus naturale* in ancient Rome. Found in the Justinian Code, the Roman *corpus civilis*, the Stoic tradition, and made popular by Cicero, this concept builds on the views of Plato and Aristotle. One of the early authors on this subject, Ulpian (third century B.C.), defines the concept of natural law by stating: "the law of nature is what nature teaches all living beings (animalia)." This view mirrors the then prevailing thought among Roman legal philosophers, that natural law was to be understood as the expressions of the laws of human nature.[11]

In the Middle Ages, a new interpretation of natural law was devised to accommodate the Church's need to survive within the reigning political systems. Thomas Aquinas developed a legal philosophy making natural law the symbol of Christian justice. The egalitarian attitude expressed in Christian dogma had originated in Judaism. The Old Testament states that its laws can be obeyed by anybody who is of good will. "The sanctification of each member of the community who obeys the law of Yahweh implies an equality of all men."[12] In this view, natural law turns out to be a general norm, positioned above the positive law, and thereby justifying appeals to the faithful to recognize the supremacy of the Church. Aquinas' concept of an unchangeable law of nature, which underscores the "eternal verities" of the true religion, was taught well into the eighteenth century.

The modern era of biology's interaction with law begins with the writings of Charles Darwin. This is a particularly interesting period because of the frequent misinterpretation of Darwin's ideas by legal scholars. The work of a contemporary of Darwin's, Sir Henry Sumner Maine, provides an example.[13] Like most legal scholars during the decades following the publication of *On the Origin of Species* (1859),[14] he used the term *evolution* without fully understanding its biological meaning. Applying his interpretation to law, he saw a natural progression of law from lower to higher states, as defined within cultures, and stated that this progression has been distinguished by the gradual dissolution of family dependency and the growth of individual obligation in its place (1885).

Modern anthropology has largely discarded Maine's theories due to the lack of confirming evidence. Studies suggest that change in social organizations, legal systems, and new legislative efforts are not necessarily steps in a progression to higher states. Rather, they are more correctly interpreted as instances of change. Another example can be found in the theories of Bachofen[15] (1815-1887), which at the end of the nineteenth century exerted a strong influence upon ethnology. Today, he is remembered mainly for his *Mutterrecht* (published in 1869) in which he challenged the prevailing patriarchal theory of social evolution. On the basis of his interpretation of evidence dealing with matrilineal descent in Greece, Africa, and America, he formulated a theory of social evolution that involved a sequence of universal stages: first promiscuity, followed by matriarchy, and then the permanent establishment of patriarchy. His theories also have been discarded because they fail to reflect available evidence.

In the United States, Darwin's theories occasionally have been embraced by some legal scholars.[16] Some American legal theorists have taken an explicitly evolutionary approach to the historical development of law. In a survey of this tradition, Elliott has shown that Oliver Wendell Holmes viewed the common law as an evolving system that changes in response to interactions between new conditions and "the deepest instincts of man."[17] Not surprisingly, this approach was associated in the public mind with ideological versions of social Darwinism. It is hardly surprising, therefore, that scientific criticisms of natural law theories should extend to literally all attempts to cross what came to be called the "naturalistic fallacy" and the "gulf" between fact and value.[18]

SOCIAL THEORIES

However ill-conceived these nineteenth century attempts at bringing evolutionary theory into legal thinking, they were not followed by a continuing influx of biological findings during subsequent decades. Toward the end of the last century, a new concept of law, *the sociology of law,* emerged among European and American legal thinkers. The concepts were promoted as an antidote to the then prevailing theories of positivism.[19] The influence of these concepts continued into this century, particularly in the writings of Emile Durkheim, Eugen Ehrlich, Max Weber, and Roscoe Pound, the latter three being important scholars in jurisprudence. Each would come to advocate a sociology of law that took into account social environment and social behavior as guidelines for legal theories.

The French anthropologist-sociologist, Durkheim,[20] proposed a developmental theory of law similar to the one that Maine had proposed earlier. For Durkheim, *division of labor* was the dominant characteristic of modern society. In his view, interdependence of the many parts of complex social orders make *contract* modern law's primary concern. Weber[21] adopted a different view, seeing *formal rationality* as the dominant characteristic of the modern legal world. In formulating his *Fundamental Principles of the Sociology of Law* just before World War I, Ehrlich[22] (1913), the Austrian legal scholar, took yet another view in introducing the concept of the *living law*, which described the interactions of people within a legal system—the law in action as opposed to the law on the books.[23]

Ehrlich's ideas are based on his observations of the law in action, that is, social interactions. Observation is the scientific tool with which ethologists build their theories of animal behavior. Scientists discern patterns in the social interactions of individuals, whether analyzing the observed behavior of humans or of other species. Ehrlich used the term *Rechtstatsachen* or *facts of law* for those patterns of human behavior that were basic facts of everyday social and legal transactions. He stipulated four facts of law: usage, domination, possession, and disposition.[24]

For Ehrlich, law is the inner order of society. Although law has many functions, its most crucial is the function of organization. The law presents the individual with alternative choices and points to those choices that the individual's society finds acceptable or just. The individual can then choose to act or refrain from acting. Knowledge of the moral consensus of the majority, which in this instance is similar to a society's concept of justice, enables an individual to predict with some degree of accuracy the consequences of a particular choice.

A second function of the law is the protection of the social order, a function supported by the decision-norm. This function complements or fills in gaps in the legal structure, thus making law more effective. Typically, the decision-norm within a society is expressed through adjudication, which enables the law to be flexible and to adapt the rigid rules of legislation to individual cases. A related point deals with laws that become obsolete or no longer reflect the views of the social majority. Rarely are such laws repealed outright. They usually continue their existence on the books for many years, even if they are not enforced. Outmoded laws can be continued or resurrected by people who are strongly bound to a particular tradition or religion, however. These traditionalists may react more strongly to the rules of their subsociety and adhere more rigorously to its laws than average law-abiding citizens.

A clear consequence of the preceding views was that law, whatever view one adopted, was no longer graven in stone. The content of the law could change with time and circumstances. Moreover, law could be used as an instrument for social change. This view accommodated the ideologically based legal theories that developed between 1900 and 1950, and that aspired to this social change. Today these ideas continue their influence, particularly where law retains its identity as an instrument of change and a tool for reform. Similar views are encountered in the ideas of modern legal philosophers, such as Rawls and Dworkin.[25] Although their philosophies differ in detail, these scholars share the belief that inequality among humans is due to social factors. Inequality is man-made, neither imposed by nature nor the will of God. It follows that legal reform with the help of man-made law can facilitate equality.[26] One may empathize with these views while still acknowledging that they lack an up-to-date understanding of the biological bases of human behavior. Missing is an understanding of concepts that connect ideas of social behavior and behavioral change to biology, and an understanding of the implications of genetic differences and of interactions between person and environment.

Social theories were not the only influences on legal theory during the early years of this century. Psychiatry and psychology also had their say. The writings of Freud and others introduced new explanations of human behavior, particularly in the areas of family and criminal law. Although these explanations remain highly influential, their validity has been questioned by recent findings in biology. For instance, work on neurotransmitters has revealed that these substances mediate specific behaviors (e.g., aggression, impulsive behavior); that physiological profiles accurately predict certain types of behavior;[27] and that one's genetic makeup largely accounts for one's personality, one's capacity to delay, and one's capacities to engage in certain types of social interactions. Such findings have been slow to find their way into legal thinking. Certainly, it takes time to make new laws, to implement regulations, and to inform those who are involved in this process. Sometimes emergencies will bring the machinery into faster motion, especially when the public feels a need for immediate action, as in the case of a growing rate of violent crime or nuclear exposure. Yet, on balance, such efforts are often haphazard, and they differ in kind and outcome from the systematic inclusion of new knowledge into legal thinking. The implications of this point are underscored by the fact that science is presently progressing at a far faster pace than ever before in human history.

An illustration of some of the preceding points is provided by premenstrual syndrome (PMS). PMS is an affliction varying in degree of severity, characterized by pain, feelings of tension, depression, anger, and unpleasant physiological changes. Approximately 20% of premenopausal adult women are estimated to suffer from this disorder. The syndrome has a biological origin, and, among a certain percentage of persons, physiological changes during the premenstrual period are associated with dramatic behavioral changes, even to the point that otherwise normally behaving women may commit physical abuse or murder. These changes in behavior occur on a predictable basis and are considered to be unintentional. Obvious questions arise concerning such issues as normalcy and responsibility for one's behavior, as well as myriad legal issues concerning methods to deal with the effects of the disorder. Legal and medical experts collaborating with scholars from related fields have been able to define a syndrome and to make specific suggestions for medical treatment and legal approaches.[28]

These include specific types of hormonal treatment, medical monitoring, and increased personal responsibility among persons who behave violently premenstrually.

Since 1950, there have been striking developments in biology that support and elucidate Darwin's theory.[29] In the 1980s, scattered attempts were made to introduce evolutionary thinking and findings into legal thinking.[30] These attempts have met with only partial success. The slow pace at which law has assimilated evolutionary ideas dealing with behavior is in part due to the intellectual isolation that is an inevitable by-product when different academic and professional disciplines define their areas of interest and expertise. In the United States the roots of this isolation can be traced to the middle of the nineteenth century when the liberal arts separated from the sciences. There have been clear consequences for biology and law: Most legal scholars failed to exploit the fundamental insights derived from biology or they often misinterpreted the implications of findings.

To argue that biologically based explanations of the human condition should be central to many of the ongoing debates dealing with legal theories and the role of law in society may at first seem like advocating continuing unproductive intellectual debate—yet such debates could result in narrowing the gap between the conflicting views. The *is* of biological reality need not be reflected in every normative prescription defining desirable conduct, to the exclusion of *ought* elements. The legal system, as it has done ever since it came into existence, will filter biological views just as it filtered Freud's theories and other influential views. The *ought* elements will eventually blend with what has been filtered. This interpretation is discussed in the writings of Bodenheimer,[31] who describes law as a bridge between *is* and *ought* and refers to the legal system as "an amalgam of is and ought elements. Law is operative in part as a living law of actual human conduct, and in part as an instrument for transforming unfulfilled social ideals or goals into reality." Given this perspective, legal ethics are not endangered if we update what we call the *is* element by replacing obsolete scientific views with newer scientific findings explaining human behavior.

ANTHROPOLOGY

For the moment, I will return to developments at the end of the nineteenth century in the field of anthropology. These developments provide a bridge into ethology. In the last years of the nineteenth century and in the early years of this century, the first results of anthropological research dealing with law and social change among primitive societies appeared. Among nearly all of the societies that were studied, relationships were reported between behavior, environment, and social rules. Malinowski's (1926)[32] studies of the Trobriand Islanders can serve as an example. His findings disproved ideas held by a number of earlier anthropologists who believed that primitive law was exclusively penal. Although such studies were important to clarify prevailing misconceptions, nearly all omitted critical findings that would affect scholarship for decades. The concept of behavior as *central* to law is an example. The concept is implicit in Malinowski's writings, yet he fails to establish that behavior is the central moving force of law. Others failed to take up the slack and, as a result, the concept has been explored infrequently since the 1920s.[33]

Nevertheless, anthropological studies in law remain helpful, primarily because they identify potential behavior-influencing distinctions such as mores, social norms, and formalized law. While these distinctions are numerous and varied, perhaps the most important one for our purposes is found in Malinowski's definition of law. As mentioned before, it includes the presence of a social machinery that can sanction deviant behavior. This distinction separates rule-following among other primates and legal behavior in humans. It does not imply, however, that man-made law can be understood free of biology. There exist rules or norms in every human society by which people live and which are rarely broken. Most of these rules and norms are not in conflict with the formal laws of the society (at least not to the extent that they are, in the eyes of the law, illegal or criminal and subject to legal prosecution). The legal machinery is seldom set in motion if any one person deviates from these norms or rules. The concept of a living law—*a law that guides the day-to-day interactions between individuals and that is expressed in individual and group behavior*—continues this element in technological societies. The most effective formal law would follow the living law. Its main functions would be to facilitate the existing flow of events, to constrain excesses, and, when needed, to restore balance. To succeed in this task legislators would not only have to be aware of the way people are inclined to act and to behave, but they would also have to be mindful of biological propensities and constraints in the human behavioral makeup.

ETHOLOGY

What is ethology? Dictionaries are not particularly helpful in answering this question and say little more than ethology is the science of character, a definition based on the use of the term dating from the last century. For an up-to-date definition one must go to ethologists themselves. There one finds a surprising degree of agreement: ethology is the biological study of behavior.[34]

Ethology is a branch of biology that emphasizes observations of animals in their natural habitats. Through direct observation, ethologists seek to identify rules of behavior vital to the survival and reproductive success of the species under study. Ethology concentrates on four main areas: ultimate causation, development, function, and proximate mechanisms. *Ultimate causation* deals with the selection factors that result in species-characteristic predispositions to behave in particular ways. *Development* addresses gene-environment interactions, which include learning. *Function* informs our understanding of how different behaviors have come to be favored by natural selection and why animals behave in predictable ways. *Proximate mechanisms* are those behavioral, physiological, and psychological events that are the initiators of short-range behavior change. An understanding of these four areas, which are the primary interests of evolutionary theory, facilitates a linking of research findings developed in various branches of biology.

Legal scholars are likely to find studies dealing with function most relevant. We speak of functions of behavior and the functions of the law, also the function of specific laws or legislation. The usage of the term differs in law and biology, however. Exploring this distinction leads to insights into numerous areas of law. Yet, an equally strong case may be made for the importance of ultimate causation. But first, more about ethology.

The term *ethology* can be traced to the writings of Sainte-Hilaire (1854) and Haeckel (1866), who used both ethology and ecology in their writings, and to the works of Espinas (1878),[35] who investigated relationships between social behavior and social order. Darwin of course was an ethologist yet the term is rarely associated with his name, nor did he use the term himself. At the turn of the century, Louis Dollo, a Belgian paleontologist, used the term in referring to his investigations into the adaptivity of behavior to the environment. Another Belgian, Emile Waxweiler, deserves special mention, for in 1906 he proposed interdisciplinary research designed to result in a synthesis of human and nonhuman social ethology. The term thus has been around for nearly a century and, since Darwin, investigators have been studying animals in their natural habitats. Yet to this day few investigators in the field of law have evinced an interest in ethology. One notable exception can be mentioned: For more than a decade a small number of political scientists have been discussing links between the behavioral aspects of social organization and their biological roots. This is the area of biopolitics, which has developed theories relating evolutionary biology, social psychology, linguistics, and game theory to politics. The field has attracted leading scholars in political science.[36]

Modern ethology came alive in the 1930s under the leadership of Karl v. Frisch, Konrad Lorenz, and Niko Tinbergen.

Their efforts led to their receiving the Nobel Prize in 1973. Although ethologists are not all of the same mind, they agree that there are important and significant genetic contributions to the behavior of each species; that during critical moments of development, gene-environment interactions influence the degree and type of genetic expression, and thus phenotype; and that behaviors and individual capacities are further modified as a consequence of both epigenesis (essentially, one's pattern of development) and learning.

Ethology is supported by rich and varied evidence, yet it is not without its detractors. Its application to humans continues to elicit strong opinions and feelings, primarily from those who insist that *Homo sapiens* is so different from all other species that comparisons with other animals are invalid or blasphemous. At times these opinions find their way into the courtroom, as in the Scopes trial in 1925, or even today when antievolutionists want equal classroom time to teach alternative interpretations of the origins of humanity. The weight of evidence clearly favors the ethologists, however, and daily the evidence becomes more compelling.[37]

Ethological findings have been the source of significant insights into human behavior, particularly when findings from other species are compared to those from humans. Cross-species similarities as well as differences are observed. In some species learning processes are limited to a specific period in an individual's life. For example, immediately after hatching, the infant greylag goose will follow its mother or any person (or dummy) supplying certain key stimuli, a process referred to as imprinting. In other species, such as mallard ducks, a sexually mature individual imprinted at birth to follow another species will likewise direct its sexual responses to that species.[38] Such findings are explained largely on the basis of the genetic characteristics of different species and the interaction of these characteristics with the timing of particular events during development.

Moving through different taxa from the most primitive to *Homo sapiens*, the degree to which genetic makeup contributes to specific behaviors declines. Thus, for example, in *Homo sapiens*, if imprinting occurs at all, it appears to be reversible. On one hand, such findings serve to underscore species differences. On the other, they underscore the importance of both genetic and environmental contributions to behavior. Applied to humans, we know, for example, that the susceptibility of numerous medical diseases (e.g., Tay-Sachs, Huntington's disease) as well as some forms of alcoholism, depression, schizophrenia, and personality (e.g., anti-social, hysterical) are influenced by genetic information.[39] Yet this is just the beginning of the list of behaviors to which humans are predisposed.

Acknowledging that genetic makeup makes a significant contribution to behavior does not mean that ethology is some form of disguised genetic determinism. Ethology also emphasizes that evidence is equally strong in demonstrating that individuals are modifiable as a consequence of different experiences and learning, and that they shape their behavior in response to cultural mores and laws. Still, there are limits to behavioral modifiability even in humans and much of human ethology is devoted to discovering and defining these limits and the processes contributing to both cross-person differences and similarities. From this perspective, the implications of biology for legal scholars and legislators are clear: Findings from the natural sciences need to be an integral part of legal thinking and decision making. Put another way: The degree to which a law will be effective is in large part determined by the degree to which the law has public support and shapes behavior within the limits of human plasticity.

THE RELEVANCE OF ETHOLOGY FOR LAW

Although there is little doubt regarding genetic influence on human behavior, there is also little doubt that individuals and groups adapt to continuously changing environments. Even though humans are not infinitely plastic, they are an integral part of the ongoing ecological process: As the environment changes, within limits set by species-characteristic constraints, the behavior of persons and groups changes. Such change may have different outcomes. These outcomes are largely determined by strategies adopted by individuals and groups. For some individuals and groups, change may lead to near extinction (e.g., Tasmanians).[40] For others, the opposite occurs. Central to this adaptive process are the rules and laws groups make and follow.

For some of the preceding points, studies of nonhuman primates result in findings that are analogous to findings from studies of humans. Such similarities imply the potential value of investigating closely related species as a means of gaining insights into human behavior. Ethologists have identified patterns of behavior among nonhuman primates that stimulate social interaction and preserve or disrupt group cohesiveness. The development of these patterns follows certain rules[41] and is influenced in part by mechanisms regulated by hormonal function. Among humans, behavior patterns such as pair-bonding and extended child-rearing predate the time when hominid ancestors separated from a common primate ancestor. Such patterns remain part of the biological heritage we share with other primates and, as such, they influence our behavior to essentially the same degree.

Similar points apply to group behavior. A reasonable assumption is that at some time during hominoid evolution our ancestors developed capacities to communicate implicit rules for group life[42] (e.g., group members should stay in close proximity, members should warn others of dangers). Rules understood by group members and without which the group could not function, but which, to the human observer's perception, are not clearly expressed are *implicit* rules of behavior, and they may be contrasted with *explicit* rules which are, to the observer, clearly expressed. With the evolution in *Homo sapiens* of capacities to speak, imagine, and plan, rules became increasingly explicit, and eventually they became formalized as customs and laws.[43]

Recent findings by molecular biologists complement the findings mentioned above and further point to a close link in the genetic endowment of apes and humans. Their techniques focus on comparing genetic material from different species, for example, by calculating the evolutionary distance of protein molecules, or by using immunological techniques, which measure antibody reactions to proteins. These techniques have placed the chimpanzee and gorilla closest to humans, with other nonhuman primates at a slightly greater distance away.[44] Such findings are relevant to the discussion in several ways. First, they call our attention to genetic contributions to behavior. Apes and humans behave differently in part because they are genetically predisposed to behave differently. Yet, both species have developed complex and elaborate communication systems. Apes are not predisposed to complex verbal language. The anatomy of their throats, which is a genetically transmitted feature, makes it impossible to formulate words. They are capable, however, of coping with complex situations and of using nonverbal messages to give directions, specify locations, and to clarify relationships. Second, such findings hint at the complexity of physiological-anatomical mechanisms underlying behavior. Third, the findings pose important questions for law. For example: Can legislators incorporate findings about gene-environment interactions into their decision making processes in ways that are beneficial to the people and societies that law serves? Or, can law provide for continuity and adaptive change in channeling human behavior faced with the continuous changes in the environment brought about by the rapid technological development over the last century?

In the following chapters, I will assume that all human behavior is shaped by the combination of several factors: genetic predispositions, developmental events, learning, and environmental contexts (e.g., social demands and options). I will also assume that certain functions (which often involve different behaviors in different environments) such as behaviors needed to develop resource access and control, or to succeed in sexual engagement with the opposite sex, are more resistant to change or regulation relative to many other types of behaviors (such as obeying specific traffic rules). It follows that deviance is likely to be particularly high where legal norms require behavior that works at cross-purposes to behaviors that are resistant to change. On the other hand, if legal norms and the behavior prescribed by such norms address those behaviors that are less resistant to change, compliance is likely to be high. Human legal behavior thus is helped or hindered by our biological nature and the demands made by the law. *The effectiveness of law will be proportional to the degree to which the function of a particular law complements the function of the behavior that the law intends to regulate.* An important objective of our inquiry, therefore, is to identify those functions of behavior that are highly resistant to change and those that are more responsive to change.

The relationship between the functions of law and behavior is not one-to-one. We know, for example, that criminal laws generally are meant to punish the guilty as well as to serve at least three other functions: to deter, to protect the innocent, and to restore social order by rehabilitation of the offender. Behaviors also often serve many functions, especially those behaviors that are important for survival. The law in its interaction with behavior often emphasizes only one of many behavioral functions and sometimes even proscribes functions that are part of a behavior. Examples of this point may be found in legislation dealing with sexuality, reproduction, and aggression. Some laws are directed to encourage or protect reproduction. Other laws will proscribe important functions of sexuality, for example sexual behavior leading to pair-bonding or survival of the family (e.g., laws against birth control and abortion). Some laws are designed to curb aggressive behavior and to protect members of society from excessive aggression as expressed in violent behavior. Other laws will allow aggression in the protection of one's own life or that of one's family. Modulated forms of aggression are accepted in efforts to earn a living and to succeed economically; unmodulated forms are not. Aggression is prescribed in times of war but not in peace time. Some behaviors can be sublimated for longer times than others. These multiple relationships between law and behavior may seem confusing or paradoxical, yet they are neither. Rather, they are the surface manifestations of a complex infrastructure in which biology and law interface.

CONCLUSION

The laws that humans obey are uniquely human. They are also deeply rooted in the biology of early humans and nonhuman primates. Laws have their roots in biology, yet they are modified by values and tradition. This point applies as much to the indigenous laws of traditional societies as to the sophisticated laws of technically advanced nations. Laws change as individuals, groups, and environments change. The time and sequence of these changes, whether law leads, follows, or lags behind, have a significant impact on the course and outcome of these changes.

NOTES

1. In this book, the terms *law, rules of law,* and legal *rules* are used in their broadest sense.

2. Gazzaniga, M. S. 1988. *Mind matters.* Boston: Houghton Mifflin. Gazzaniga describes the relationship between mind and brain as follows: "The mind is in some kind of dynamic equilibrium with the cellular and network aspects of the brain."

See also Gazzaniga, M. S. 1985. *The social brain.* New York: Basic Books.

3. The structure and size of early human organization has been documented by archaeological data (e.g., Leakey, L., and V. Goodall. 1969. *Unveiling man's origins.* Cambridge, MA: Schenkman) as well as studies of present-day hunter-gatherer societies (e.g, Eibl-Eibesfeldt, I. 1972. *Die !Ko-Buschmann-Gesellschaft.* Munich, Germany: R. Piper) and by studies of nonhuman primates (e.g. Kummer, H. 1971. *Primate societies.* New York: Aldine Atherton).

4. Lieberman, P. 1984. *The biology and evolution of language.* Cambridge, MA: Harvard University Press.

5. Fox, R. 1983. *Red lamp of incest.* South Bend, IN: University of Notre Dame Press.

6. Tiger, L. 1988. *The manufacture of evil.* New York: Harper & Row.

7. Malinowski, B. [1926] 1972. *Crime and custom in savage society.* Totowa, NJ: Littlefield, Adams.

8. Of course, it is hypocritical to blame only those who profited from this situation. In the consumption of alcohol, or any illegal behavior (e.g., prostitution), it takes two to tango. Al Capone said: "I make my money by supplying a public need. If I break the law, my customers, who number hundreds of the best people in Chicago, are as guilty as I am. The only difference between us is that I sell and they buy." (Kyrig, D. E., ed. 1985.*Law, alcohol and order: Perspectives on national prohibition.* Westport, CT: Greenwood.) Here are some statistics describing the effects of the Prohibition Act in the United States: between 1918

and 1922 imports in liquor increased: from Canada, 6 times; Mexico, 8 times; West Indies, almost 5 times. Imports from the Bahamas increased from less than 1,000 gallons to 386,000; Bermuda, from less than 1,000 gallons to 40,000. A low estimate of the value of imported (illegal) liquor puts it at $20 million in 1922, $30 million in 1923, and $40 million in 1924. The toll of death and illnesses due to the consumption of poisonous homemade alcoholic beverages ran into the thousands per year. The sale of grapes to produce alcoholic beverages doubled between 1920 and 1926. According to a 1932 estimate, the consumption of wine rose by two-thirds during prohibition. Obviously, the number of people consuming alcohol and thereby committing a criminal act was large. Efforts to enforce the law could not keep up, although the number of persons in prison increased from 32.3 per 100,000 in 1919 to 34.6 in 1923 and 41.8 in 1926.

The direct result of this legislation was the emergence of powerful organized crime run by the Italian Mafia, which started in Chicago. Organized and syndicated crime had been controlled by existing established Jewish and Irish gangs until such time. The underworld already existed, but the newly found opportunity to make tremendous profits by the sale of illegal alcoholic beverages gave it a tremendous boost. Loan-sharking increased dramatically because the enormous profits generated from bootlegging were available to loan. Labor and business racketeering, another source of profits, began in the 1920s and in the next 10 years became a major source of syndicated income. Organized crime became so powerful that after the repeal of the Act in 1933 it continued to use this power in profiting from other human desires for illegal behavior, as for example prostitution and gambling. At the same time, it organized the import and the sale of drugs that is now the main source of racketeering profits in this and many other countries. It is estimated that more than 90% of criminal cases brought to trial in this country to date are drug connected in some way. (Cashman, S. D. 1981. *Prohibition: The lie of the land*. New York: Free Press.)

9. Masters, R. D. 1977. Nature, human nature, and political thought. In *Human nature in politics*, eds. R. Pennock and J. Chapman, 69-110. New York: New York University Press.

10. The classic statements are in Book IV of Plato's *Republic* and Book VII of Aristotle's *Ethics*. For a more extensive treatment, see Strauss, L. 1953. *Natural right and history*. Chicago: University of Chicago Press.

11. Friedrich, C. J. 1969. *The philosophy of law in historical perspective*. Chicago: University of Chicago Press.

12. Ibid.

13. Maine, H. 1861. *Ancient law*. London: J. Murray.

14. Darwin, C. [1859] 1964. *On the origin of species*. New York: Atheneum.

15. Bachofen, J. J. (1815-1887), a contemporary of Charles Darwin (1809-1882), based his ideas on a theory of social development that maintains that the first period of human history was matriarchal. In his main work, titled *Mutterrecht und Urreligion* (1869), and in other essays, he intended to discover the universal law of history. He was a jurist and a historian of Roman law. In his historical view he saw legal concepts, such as homicide, revealing profound changes in man's appraisal of human life, of shifts in social organization, of reinterpretations of social classes, of growing egalitarianism, of the emancipation of women, of modifications of technology. All this he accepted as obvious; what he wanted to discover was what accounts for these changes.

Bachofen concluded that legal concepts are interrelated with religious, ethical, and aesthetic beliefs. Ironically, he is mainly remembered in his role of inspiring some hypotheses by Karl Marx and Friedrich Engels. They used a small part of his writing, in which he discussed a supposed primal stage of communal sexuality, to support their hypotheses of an early communal order of civilization.

(*Myth, religion, and mother right: Selected writings of Johann Jakob Bachofen,* trans. R. Manheim. 1967. Bollingen Series No. 84. Princeton, NJ: Princeton University Press.)

16. See Hofstader, R. 1955. *Social Darwinism in American thought* rev. ed. Boston: Beacon Press. For a classic example, see Sumner, W. G. 1963. In *Social Darwinism: Selected essays of W. G. Sumner,* ed. Stow Pearsons. Englewood Cliffs, NJ: Prentice Hall.

17. Elliott, E. D. 1985. The evolutionary tradition in jurisprudence. *Columbia Law Review,* 85:1

18. Brecht, A. 1959. *Political theory.* Princeton, NJ: Princeton University Press.

19. Supra note 17.

20. Durkheim, E. 1933. *The division of labor in society.* Trans. G. Simpson. Glencoe, IL: Free Press.

21. On the concept of rationality in Weber's work, see Parsons, T. 1964. Introduction. In Max Weber, *The theory of social and economic organization;* and Rheinstein, M. 1954. *Max Weber on law in economy and society.* Cambridge, MA: Harvard University Press.

22. Ehrlich, E. [1913] 1975. *Grundlegung der Soziologie des Rechts.* Translated as *Fundamental principles of the sociology of law.* New York: Arno.

23. Rehbinder, M. 1989. *Rechtssoziologie.* Berlin: Walter de Gruyter; Podgorecki, A. 1981. Living law. *Law & Society Rev.* 15:1.

24. Gruter, M. 1982. Biologically based behavioral research and the facts of law. *J. Soc. Biol. Struct* 5:315-323.

25. Dworkin, R. 1977. *Taking rights seriously.* Cambridge, MA: Harvard University Press; Rawls, J. 1971. *A theory of justice.* Cambridge, MA: The Belknap Press of Harvard University Press.

26. For a detailed discussion of present-day sociology of law, see supra note 23.

Rehbinder's revised edition of *Rechtssoziologie* (1989) suggests that the development of a *legal ethos* (Rechtsethos) has to complement the empirical work on a general sense of justice, the attitudes to judicial institutions and authorities, or prestige of the law. He emphasizes the inexactness of predicting individual behavior.

27. McGuire, M. T., and M. Raleigh. 1988. *Legal concepts in the face of research on the neurochemical substrates of social behavior,* unpublished manuscript.

28. Ginsburg, B., and F. Carter, eds. 1987. *Premenstrual syndrome.* New York: Plenum.

29. Wilson, E. O. 1975. *Sociobiology.* Cambridge, MA: Harvard University Press.

30. Elliott, supra note 17; Epstein, R. 1980. A taste for privacy? Evolution and the emergence of a naturalistic ethic. *Journal of Legal Studies* 9:665; Gruter, M. 1976. Die Bedeutung der Verhaltensforschung für die Rechtswissenschaft. In

Schriftenreihe zur Rechtssoziologie und Rechtstatsachenforschung, Bd. 36, eds. E. E. Hirsch and M. Rehbinder. Berlin: Duncker & Humblot; Gruter, M. 1977. Law in sociobiological perspective. *Florida State University Law Review* 5:2; Gruter, M., and P. Bohannan. 1983. eds. *Law, biology and culture: The evolution of law*. Santa Barbara, CA: Ross-Erikson; Gruter, M., and R. D. Masters, eds. 1986. *Ostracism: A social and biological phenomenon*. New York: Elsevier Science; Hirshleifer, J. 1980. Privacy: Its origin, function and future. *J. Legal Stud.* 9:649; Rodgers, W. 1982. Bringing people back: Toward a comprehensive theory of taking in natural resources law. *Ecology L.Q* 10:205.

31. Bodenheimer, E. 1988. Law as a bridge between is and ought. *Ratio Juris* 1:2.

32. Malinowski, supra note 7.

33. E. A. Hoebel, in his *Law of primitive man* (1954, Cambridge, MA: Harvard University Press), speaks of law as behavior, but he does not elaborate on the idea. There are, of course, limitations to viewing law only from the perspective of behavior. P. Bohannan, for example, points out that in observations of behavior, observers will color what they see. This is one of the main difficulties in studying legal behavior: our feelings and value judgments influence our interpretations. "The concrete system necessarily remains a mystery. All we can do is to learn more about our sensory means of perception, and mechanical and other extrinsic extensions of them, and our cultural prison." (Bohannan, P. 1969. Ethnography and comparison in legal anthropology. In *Law in culture and society*, ed. L. Nader. Chicago: Aldine.)

In his new book *We the alien* (in press), Bohannan starts the introduction with a statement: "The emphasis of this book is that human behavior is animal behavior. However, all human behavior is also culturized behavior. Neither view alone can be satisfactory, human behavior is shaped by both biology and culture."

34. Slater, P. J. B. 1985. *An introduction to ethology*. Cambridge, UK: Cambridge University Press.

35. Espinas, A. 1878. *Über die tierischen Societäten*. Paris, France: Bailliere; see Wickler, W. 1972. *The sexual code*. Garden City, NY: Doubleday.

36. See, for example, Masters, R. D. 1989. *The nature of politics*. New Haven, CT: Yale University Press.

37. Tinbergen, N. 1974. Ethologie. In *Kritik der Verhaltensforschung: Konrad Lorenz und seine Schule*, ed. G. Roth. Munich: C. H. Beck.

38. Lorenz, K. 1965. *Evolution and modification of behavior*. Chicago: University of Chicago Press; Lorenz, K. 1952. *King Solomon's ring*. New York: Y. Crowell.

39. Plomin, R. 1990. The role of inheritance in behavior. *Science* 248:117-122.

40. *Encyclopaedia Britannica*, 15th ed., s.v. "Tasmanians."

41. Smuts, B. B., et al., eds. 1987 *Primate societies*. Chicago: University of Chicago Press.

42. Lewis, D. B., and D. M. Gower. 1980. *Biology of communication*. New York: John Wiley.

43. Else, J. G., and P. C. Lee, eds. 1986. *Primate ontogeny, cognition and social behavior*. Cambridge, UK: Cambridge University Press.

44. Edey, M. 1972. *The missing link: The emergence of man*. New York: Time-Life Books.

TWO

Selected Theories and Findings from Biology Relevant to Law

When the behavior of *Homo sapiens* is compared with that of other species, two points are clear: hominoid evolution covers a period of several million years; and even the most preliterate human societies are highly advanced relative to our closest phylogenetic relatives, the nonhuman primates. Nevertheless, we share many features with nonhuman primates, and studies of their behavior and physiology are likely to provide important insights into human behavior.

Perhaps an appropriate place to start the discussion of biological findings relevant to the law is with comments about the meaning and usage of the term *biology* and the biological approach to behavior. Inside the discipline of biology, the term biology is used almost exclusively to refer to the science or discipline that studies life. Outside biology (e.g., medicine, philosophy) the term typically references physiology or genetics.[1] I will use the term as biologists do, which references the study of life and life forms, two features of which are physiology and genetics. Considering the biological approach to behavior, biologists generally agree that species evolution is characterized by the natural selection of behaviors that are associated with differential survival and reproductive success. Behaviors not associated with high rates of survival or reproductive success become the waste products of evolution and, on average, appear less often in succeeding generations. Basically, there are two ways successful behaviors may be transmitted from generation to generation. One is via genetic information. Animals are born with predispositions to behave in certain ways. Withdrawal from a painful stimulus is an example. No one needs to be taught to behave this way. A second way is by experience and learning. Learning, however, does not occur in a vacuum, even among *Homo sapiens*, who has a more versatile learning system than other species. The capacity to learn also has been selected. Generally, most humans can learn some tasks more easily than others due to the fact that their genetic makeup has primed them for such learning. Nearly all normally developed humans learn to walk upright or to avoid an obstacle in their path, as well as to speak and count. There are also constraints to the types of behaviors that can be learned and performed. Playing the piano, for instance, is not a behavior that has been selected. Thus, the piano is difficult to learn and only a few individuals play to perfection. Or, most humans probably cannot learn most bird calls no matter how hard they try; and certainly we are not built to fly. Constraints that prevent us from behaving in certain ways represent evolutionary outcomes just as much as do predispositions to behave in certain ways. The concept of constraints thus needs to enter our inquiry, and this inquiry will reveal that one of the main functions of man-made law is to strengthen constraints.

More focused investigations of biologists deal with such phenomena as neurotransmitters, receptor sites, and ion channels. These are the various subparts of neuronal function that are responsible for operationalization of species-characteristic behaviors, such as developing affectional bonds and competitive behavior. Short of extreme conditions (e.g., social isolation or multiple placements during early childhood), predisposed behaviors contribute significantly to an adult's phenotype, a point underscored by the similarities in human behavior across different cultures.[2]

Biologists also use guidelines for extrapolating from nonhuman to human behavior. The first of these deals with analogies and homologies. *Analogy* references cross-species similarities in functions. For example, mother-infant bonds are seen in nonhuman primates, carnivores, whales, birds, and humans. In each of these species they serve essentially the same function, that of providing infants with both nutrients and protection. *Homology* references cross-species similarity in terms of evolutionary history. Functional similarity may or may not be present. The wings of birds and the arms of humans are homologous, yet they have come to serve different functions. Analogous relationships are of more immediate interest to investigators because they provide hints as to how different species have evolved in solving critical problems of living. A second guideline concerns differences between ultimate and proximate explanations of behavior. *Ultimate explanations* reference why certain behaviors were selected (i.e., favored in preference to other behaviors during evolution). *Proximate explanations* reference the details of current behavior, for example, what neurotransmitter changes occur in response to being frightened by a snake. Proximate mechanisms may also have ultimate causes, but the investigative focus is usually on their short-term operation, not their evolutionary history. We will be concerned with both ultimate and proximate explanations. Law often assumes ultimate explanations and focuses on proximate explanations. Ultimate explanations, however, can inform law in many important areas.

A third guideline deals with the interpretation of data from other primate species, particularly in two areas: frequency of observation and conditions of observations. In some instances, the behavior of animals has been observed only occasionally and inadequate data serve as the basis for extrapolations to humans. This is a questionable practice. Even more questionable are generalizations based on observations made only of captive animals. Incarceration (externally imposed limitations on behavior) generally results in behavior changes not seen in natural settings. Misinterpretations of behavior often follow. (Scientists are aware of this issue, and in many animal studies this phenomenon is turned into a basic tool of research, as in instances where changes in blood levels of hormones are studied in response to incarceration.) An obvious consequence of such misinterpretations is that cross-species comparisons (the comparative study of animal behavior) can become confusing, irrespective of whether a writer has a positive or negative attitude to this field of science. A fourth guideline is a warning, namely, to be skeptical. Biologists like to ground their theories in evidence, and especially so when they are comparing two or more species. Generally, they try to avoid making generalizations or inferences that they cannot defend with evidence. These precautionary measures are designed to minimize misinterpretations of findings and the uncritical acceptance of scientific theories. The case of social Darwinism, which combined a misreading of Darwin's theories and an uncritical acceptance of available naturalistic data on humans, illustrates this point. An extrapolation of this point has to do with simplifying science: Misinterpretation of biological research, when brought into a legal framework, can result when single hypotheses requiring further investigation are developed as opposed to sets of hypotheses (see Alexander's commentary on Beckstrom[3]).

KIN SELECTION—INCLUSIVE FITNESS—NEPOTISM

Taking the preceding points as background, we can look at selected examples of cross-species similarities in anatomy, physiology, behavior, and, in some instances, cognitive function. How might some of the findings from animal behavior apply to law? Studies and theories of altruism can serve as examples. During the last 20 years, several theories have been developed based on this concept. Animals can exhibit behavior that appears to be altruistic. Lorenz[4] referred to such behavior as "moral-analogous behavior in social animals." In discussing kin-related altruism Wickler noted: "Helping one's kin can insure the survival of one's own genetic material."[5] In effect, in striving for one's own or one's kin's reproductive success an individual attempts to insure the survival in subsequent generations of the material carried in his or her genes. The explanations discussed below were developed to explain behavior (intentional or unintentional) that, at initial reading, appears to be incompatible with the idea that individuals act primarily in their own self-interest. Altruists, or persons who sacrificed resources and energies by investing in (helping) others, appeared to be disadvantaged in terms of genetic replication. It followed that genes predisposing a person to altruistic behavior would be selected against and gradually become extinct. Biologists thus were at a loss to explain the existence of altruistic behavior in both human and other species. The two biologists best known for resolving this dilemma and explaining altruism in biological terms are Hamilton[6] and Trivers,[7] who, respectively, developed the *kin-selection* and *reciprocal altruism* theories. I will deal with these in order.

Kin-selection theory builds from our knowledge that proportions of genetic material shared by individuals vary depending on their degree of relatedness. Hamilton took this idea as the basis of his explanation of altruism among kin: In effect, unidirectional investment in kin should increase as the proportion of relatedness between investor and recipient increases; the net effect of such investment is an increased probability of genetic replication in the next generation. Put another way, kin-selection theory explains why I will invest in close kin (e.g., offspring, siblings, nieces) who are likely to reproduce and why I will often do so *without* expecting to be reciprocated. I am investing in future replicates of myself.

Preferential investment in kin should operate when, on average, the loss in fitness to the person who gives is potentially more than offset by an increase in the giver's inclusive fitness (a measure of the giver's genetic replication in subsequent generations). Such behaviors can be both selfish and altruistic at the same time. Investment in kin, for example, is altruistic because it involves a cost to the investor, but selfish because it represents an investment in the investor's genes. From an evolutionary perspective, the theory explains selection of genes leading to behavior that increases the fitness (survival and reproductive potential) of genetic relatives even at some cost to the individual who is investing. Evidence from a variety of species, especially among nonhuman primates and many mammals, support predictions made from the theory. Nonhuman primate parents, for example, protect their own and the offspring of their siblings more than the offspring of unrelated animals living in the same group. Among humans, the phenomenon of favoring and investing in kin is well known—we call it nepotism. (The German colloquial term *Vetterles Wirtschaft* describes it more vividly.)

INHERITANCE LAW

Favorable treatment of family members and special rules for giving to kin are universal. Generally, the law recognizes this special relationship. Moreover, family bonds are considered to be vital factors in the maintenance of group and social order. An understanding of these points may explain our tolerance for preferential investment in relatives. Such behavior is recognized almost universally by the law in areas dealing with inheritance and preferential rights. In most social groups, the laws of inheritance favor descendants of the deceased, with some provisions being made to allocate a share of the estate to a surviving spouse (if applicable). For example, the Chart of Relationships Through a Common Ancestor (see Figure 2.1) is the guideline used by the California Bar to allocate the assets of the estate, if a person dies without a valid will or testament—what the law calls intestate succession. In other states, the Table of Consanguinity (see Figure 2.2) is generally the basis of statutes dealing with intestate succession.

In the United States, these statutes differ only slightly from state to state and then mainly with respect to the share of the surviving spouse. In both charts, the allocation of the right of succession follows relatively clearly from calculations of the degree of genetic inheritance. Thus if there are no children or

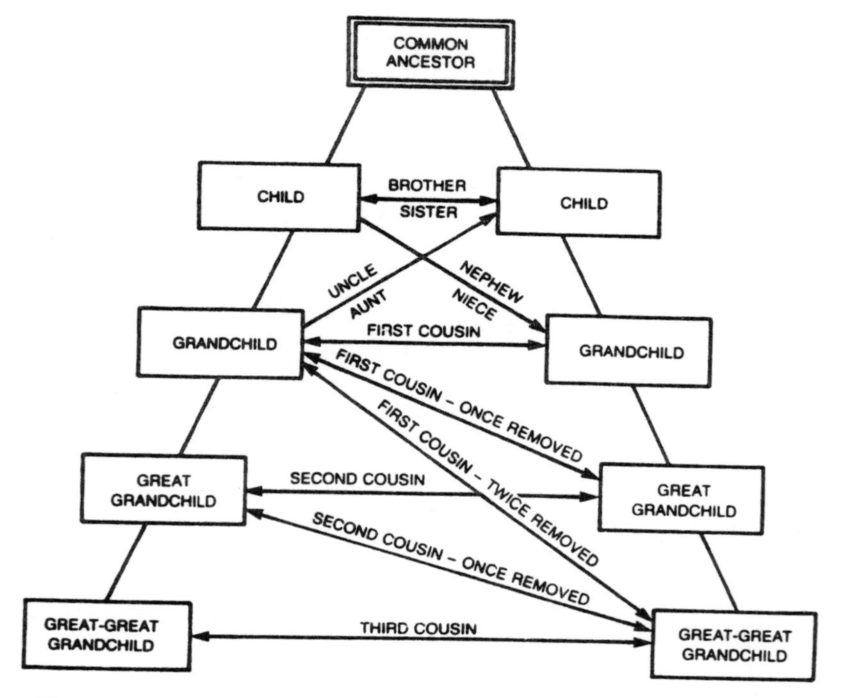

Figure 2.1. Chart of Relationships Through a Common Ancestor
SOURCE: © 1960, 1990 by The Regents of the University of California. Reproduced with permission from California Continuing Education of the Bar's 1960 book, *California Estate Administration*.

offspring of children surviving, siblings or their offspring will inherit. Where these charts are inadequate from a kin-selection perspective is in situations where there are illegitimate children. Rarely have their rights been protected. The development of legal rules for adoption opened up yet another set of possible family-bonding relationships. In these cases, many of the rights of genetic kin are given to non-kin.[8]

The point here is not that wills and intestate succession laws reflect their framers' knowledge of genetic relatedness and transmission. Rather, it is that the statutory provisions in a heterogeneous country like the United States, where there are 50 different independent states regulating succession by statutes, are strikingly similar and illustrate how closely emotional

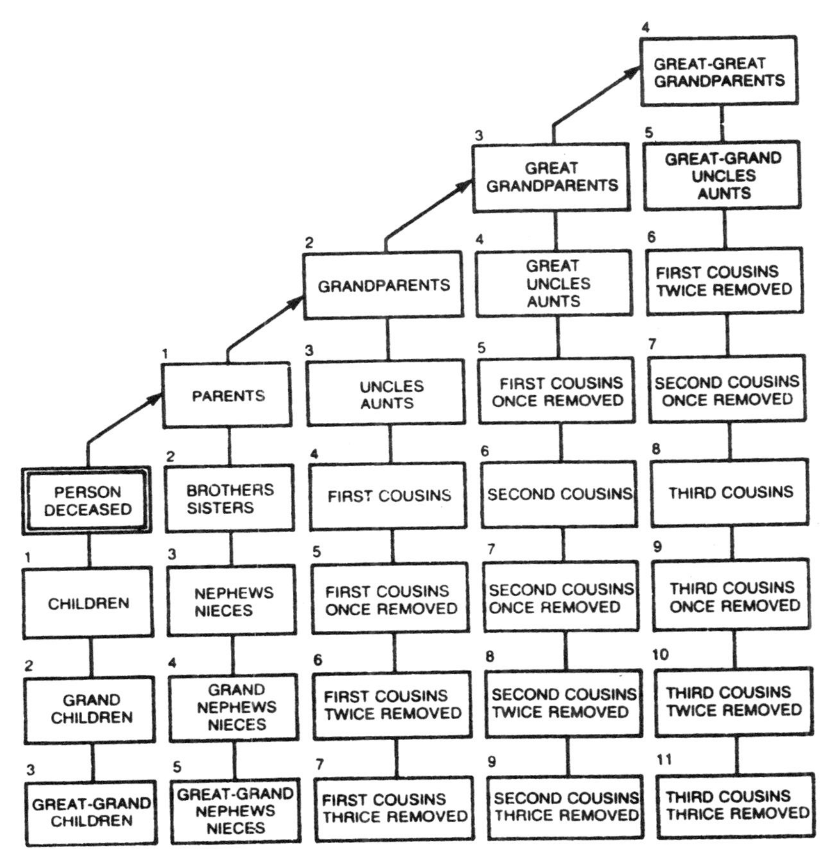

Figure 2.2. Table of Consanguinity (Showing Degrees of Relationship).
SOURCE: © 1960, 1990 by The Regents of the University of California. Reproduced with permission from California Continuing Education of the Bar's 1960 book, *California Estate Administration*.
NOTE: Figures show degree of relationship.

feelings and rational decisions follow predictions from kin-selection theory in the area of estate and inheritance.[9]

Legislation frequently grants exceptional rights or duties on the basis of kinship. Western law abounds with regulations that facilitate transfer of a person's estate to his or her descendants. Often these methods result in favored tax transfer, thus leaving a higher share to descendants. For instance, in Germany the amount of inheritance taxes levied on the estate depends on the relationship of the individual heirs to the deceased. Direct descendants pay lower taxes than more distant relatives or strangers. In the United States, the use of trusts to accomplish the goal of lowering estate taxes has generally met with public support. Spouses have not always fared as well as descendants in Western inheritance laws, however. Recent legislative trends in the United States, which emphasize community property, are particularly interesting with regard to this point. New laws strengthen the spouse's inheritance rights.[10] These changes would be predicted by kin-selection theory if the deceased leaves minor children: namely, support of the spouse with child-rearing responsibilities will benefit the deceased spouse's kin—his or her children.[11]

Kinship-related preferential rights are not limited to inheritance and succession statutes. They also influence legislation in such areas as maternity leaves, employment of more than one member of a family in an institution, and limited employment of children. Many laws are enacted with the assumption that there will be preferential giving to kin, such as mothers feeding infants. Laws based on this assumption usually are designed to enact penalties if the assumed behavior does not occur (e.g., failure to feed one's infant). Of particular interest is the degree to which laws might be enacted that are in conflict with the tendency to invest in kin. Nepotism is part of many accepted social mores. We raise our children, pay for their college, baby-sit our grandchildren, and so on. Helping kin, employing one's own offspring or relatives, or favoring them socially are commonly accepted practices. In larger groups or work situations, nepotism often results in a form of favoritism that is disadvantageous to nonrelated employees, hence, the frequency of laws or institutional rules that constrain relatives from working together.[12] In other areas, laws make other distinctions based on kinship. The selective transmission of "confidential" information concerning business matters would be expected among kin on the basis of kin-selection theory. Moreover, the theory would predict that it is extremely doubtful that even the most stringent laws will prevent it.

RECIPROCAL ALTRUISM

Let us now turn to altruism among non-kin, behavior that is explained by *reciprocal altruism theory.* This theory provides an explanation as to why non-kin engage in reciprocal helping behavior. The idea is as follows: A invests in (e.g., assists, helps) B because, at some future time, B will reciprocate A's assistance. From the standpoint of A, the ideal strategy would be to have the cost of helping B be less than the benefit accrued to A by B's reciprocation. In this way, A would gain from B. The theory thus explains altruistic behavior (A investing in B) while at the same time explaining selfish behavior (A helps B so that B will help A in return). Moreover, the theory provides an explanation for why such behavior would be favored by selection: To the degree that the benefits received from helping equal or exceed the costs of helping, and such benefits contribute to increased reproductive success, helping behavior should be selected.

Reciprocal relationships have been demonstrated in bats, carnivores, some mammals, birds, and humans.[13] This fact suggests that such relationships have a long phylogenetic history and that humans are likely to be strongly predisposed to such behavior.

The implications of reciprocal altruism theory for law are multiple and range from the need to constrain tendencies to reciprocate (antitrust legislation), to setting limits on information flow in the stock market, to establishing procedures for sealed bids for business contracts. The strength of reciprocal relationships has important implications for other legislative and enforcement actions as well, yet there is a paradox here. On one hand, it is recognized that a large degree of social order and behavioral predictability, be it the tendency of most drivers to obey stoplights or the implicit tendencies among younger persons to help elderly neighbors in distress, is based on the assumption that people tend to engage in helping or reciprocal behavior. On the other hand, laws that restrict reciprocity (such as those restricting the exchange of privileged information) are likely to be met with a high degree of noncompliance, no matter what degree of enforcement. (Life often seems full of pitfalls: Those who make laws designed to constrain abusive tendencies of reciprocal behavior often need to engage in reciprocal behavior with fellow legislators to get the laws passed.)

Both kin-selection and reciprocal altruism theory can serve to inform our understanding of group behavior. Family bonds and kinship are essential for maintaining group order. Order is further maintained by within-group hierarchical relationships. An adult who is the head of the family not only assumes various roles (e.g., mother, provider, companion), but also assigns roles to other family members depending on their age, sex, or ability. A consequence of this role assignment within kinship groups is the redirection of some motivations from individual to group, from pure selfishness to selfishness combined with altruistic actions, such as family members providing for kin who are ill. Among groups composed of non-kin, the theory of reciprocal altruism provides a similar explanation of cooperative behavior.

It is but a short step from investment in others to the development of interdependent groups of related and nonrelated individuals and the need to develop group-relevant rules. This point is particularly true as groups increase in size and the degree of interdependence increases among members. Altruistic traits, the realization that one must often depend on non-kin, and an affinity to seek a place within a structured group thus are likely to have been major contributing factors in the evolution and refinement of rule-making and rule-following behavior.

The evolution of rule-making capacities as well as the desirability of making rules are likely to have enabled some of our ancestors to shift from their hunter-gatherer existence to a life-style supported by agriculture (ca. 10,000 years ago).[14] Another shift occurred in association with the industrial revolution. Groups gradually became so large, and so many members were unrelated, that kin relationships alone were no longer sufficient to guide interactions. Cooperation and altruism within urban areas and among complex nation-states became the only alternative compatible with survival.

Kin-selection theory and reciprocal altruism theory explain general predispositions. More detailed biological information is also of potential interest to the law. The following example, which illustrates this point, focuses on the function of the neurotransmitter *serotonin.* Neurotransmitters are chemicals that communicate information between neurons in the brain. For more than a decade it has been known that reduced serotonin activity in the brain is associated with impulsive and aggressive behaviors in certain persons. The relationship between serotonin activity and aggressive behavior was known in nonhuman primates in the 1970s.[15] In 1988, Roy and Linnoila[16] reported on a group of convicted males who had known histories of arson and impulsive-aggressive behavior. Eighty percent (80%) of the subjects who had low serotonin activity committed the same type of crime within the subsequent 3 years. Decreased serotonin activity has been shown to be associated with early-onset alcoholism, a disorder that is both distinguished from late-onset alcoholism and associated with an increased probability of impulsivity and aggression.[17]

The relationships between serotonin and violent behavior suggest a number of interesting legal questions. For example, should assessments of brain serotonin activity be required legally for persons with a history of violent or impulsive behavior? If brain serotonin activity is found to be low, should treatment be required? Should persons with low serotonin activity be viewed primarily from a legal perspective or should medical considerations be introduced?

CONCLUSION

In this chapter, I have discussed the use of the term *biology*, biological concepts and reasoning, and reviewed two important biological theories—kin-selection and reciprocal altruism—both of which have clear implications for law. These theories set the stage for a more detailed examination of legal behavior among groups, starting with relevant findings in families and continuing through to the behavior of large groups. This examination is the subject of the next two chapters. In addition, I have reviewed some specific biological data relating physiological characteristics and violent behavior in order to illustrate how detailed biological information can raise specific questions about behavior that the law needs to address.

NOTES

1. Alexander, R. 1986. Biology and law. In Gruter and Masters, eds. *Ostracism*.

2. Eibl-Eibesfeldt, I. 1984. *Die Biologie des menschlichen Verhaltens*. Munich, Germany: R. Piper.

3. Alexander, supra note 1.

4. Lorenz, supra note 38, Chapter 1.

5. Wickler, supra note 35, Chapter 1.

6. Hamilton, W. D. 1964. The genetical evolution of social behavior. *J. Theor. Biol* 7:1-16.

7. Trivers, R. 1971. The evolution of reciprocal altruism. *Quarterly Review of Biology* 46.

8. Historical records indicate that adoption was originally invented as a legal substitute for kinship in early Western civilizations, mainly to transfer wealth and status. *The law of Gortyn* (Bücheler, F. and E. Zitelmann. [1885] 1974. *Das Recht von Gortyn*) devotes a considerable part of the encoded text to this institution.

9. This argument differs from that made by Beckstrom. He contends that intestate succession laws should reflect subconscious wishes of persons dying without a will and assumes that in this situation the degree of genetic relatedness will influence a dying person's wishes. Beckstrom, J. H. 1985. *Sociobiology and the law*. Chicago: University of Illinois Press.

10. The move from personal property to community property, supported by feminist movements in various states during the 1970s and 1980s, was intended to and in general has improved the spouse's position in this situation (*12A Am Jur 2d, Community Property §§5,6*). Support of the spouse is not in contradiction to kin-selection theory as long as the spouse cares for the children

of the deceased, which, again in evolutionary terms, has been the most common situation through history. Miller Weisberger, J. 1986. Marital property discrimination: Reform for legally excluded women. In *Ostracism*, eds. Gruter and Masters.

11. In civil law, child support is a legal obligation of both father and mother in all states of the United States. The relationship of parent and child and the character and extent of parental obligation, including the mother's right to reimbursement from the father, is spelled out in detail in state and federal law (*59 Am Jur 2d Parent & Child §42 §53*). The criminal law deals with desertion and nonsupport. The protection of the law extends to unborn children, adopted children, illegitimate children, and children conceived by artificial insemination (*23 Am Jur 2d Desertion and Nonsupport §50 to §64*). The California Penal Code imposes imprisonment and fines on parents of children under the age of 14 years when found guilty of desertion with intent to abandon. Similarly the law deals with neglect of children under the age of 14 (*271 West's Annotated CA Codes Penal Code*).

The law has protected children under the age of 14. Certain parental duties have always been required by law as long as persons were minors. In California, the age limit for minors has been reduced to the age of 18. In some cases the law grants the status of "emancipated minor" to persons under the age of 18. Originally the law was meant to protect a minor's rights against interference by the minor's parents in his or her dispositions concerning health, property, and similar matters. This law is now also being used by parents who wish to limit their responsibilities toward a child in certain situations (*Calif. Civil Code §62 Minor*).

12. The term *Vetterles Wirtschaft* is used colloquially in Germany to indicate that reciprocal helping arrangements within families (*Vetter* means cousin) often prevent others from enjoying equal access to favored positions.

13. Taylor, C. E., and M. T. McGuire. 1988. Reciprocal altruism: 15 years later. *Ethology and Sociobiology* 9:2-4.

14. Wills, W. H. 1988. *Early prehistoric agriculture in the American southwest.* Santa Fe, NM: School of American Research.

15. Raleigh, M., and M. T. McGuire. 1980. Biosocial pharmacology. *Journal McLean Hospital* 2:73-84.

16. Roy, A., and M. Linnoila. 1988. Suicidal behavior, impulsiveness and serotonin. *Acta Psychiat. Scand* 78:529-535.

17. Buydens-Branchey, L., M. H. Branchey, D. Noumair, et al. Age of alcoholic onset. II. Relationship of susceptibility to serotonin precursor availability. *Arch. Gen. Psychiat* 46:231-236.

THREE

The Origins of Social Organization: Kinship Bonds and Family Structure in the Animal World

Although humankind alone writes laws to influence social and family organizations, we share with other species many of the behavioral propensities and brain mechanisms that lead to the origins of social organization, among them family formation and intrafamily interactions. In Chapter 2, biological explanations for such behavior were introduced in the review of the theories of kin-selection and reciprocal altruism. In this chapter, evidence explained by these theories is examined in greater detail. The primary focus will be on the behavior of nonhuman primates.

Like humans, nonhuman primates exhibit wide diversities of family structure. Among chimpanzees, the family unit consists primarily of a mother and her offspring. Among baboons, several different structures are observed. One adult male can bond with one female or with several females to form a polygynous group. A closer look at baboons will be instructive not only for gaining insights into the building blocks of primate family structures but also because the inquiry illustrates how ethology can inform law. The behavior of hamadryas baboons will be discussed first.

The average hamadryas baboon troop comprises approximately 40 animals, with 7 to 9 adult males. The males weigh approximately twice as much as the females, have dagger-like teeth, and show aggressive behavior from early infancy. Adult males protect the troop: Any three can successfully defend the troop against all predators except a pride of lions.[1]

Hamadryas baboons live in polygynous family units (composed of one adult male, several females, and their offspring). The male forms strong pair-bonds with the females. These one-male units have been extensively studied in the field by Kummer, who has suggested that the composition of such units represents an adaptive solution to the particular characteristics of the environment in which hamadryas baboons live.[2] The size of the units appears to be optimal for foraging for this species: They are large enough to include one male protector for several females and their offspring, yet small enough to facilitate procurement of sufficient food for all.

Hamadryas pair-bonds are in part regulated and maintained by *herding*, a term that describes the behavior of adult males who actively constrain and direct the behavior of females. The larger size of males makes herding relatively easy. A bond with one female does not appear to disturb bonds with other females in which the same male participates. Hamadryas females may be associated with a particular male for a number of breeding seasons. Infants and small juveniles are bonded to their mothers. Thus a larger band is divided into subunits along familial lines. This type of group structure is seen across a number of nonhuman primate species.

PAIR-BONDS—PERCEPTION OF BONDING—RESPECT FOR POSSESSION

The hamadryas males' inclinations to herd may be effective in fostering family bonds in part because females respond submissively when they are approached by displaying males who are herding. In similar situations, females of most other primate species take flight. (Female chimps, however, seem to learn during adolescence correctly to interpret and respond to male displays that may constitute either threatening or courtship behavior.) Hamadryas males are possessive of the females in their units and particularly so in relationship to reproduction. Moreover, the intensity of male possessiveness is strikingly similar across units, a point that is underscored by findings from a colony of hamadryas baboons located at the Russian Research Station at Sukhumi. After several generations of captivity, adult males still showed herding behavior. "Apparently, even a drastic change of environment does not necessarily alter the herding syndrome."[3] The strength of this possessiveness and the associated aggression is described in a report from Kummer:

> When 30 new females were introduced to a colony of about 100 hamadryas baboons in the London Zoo, all the old males tried to secure females and, within one month, killed 15 of them in competitive fights over their possession. This event, though provoked by an unnatural manipulation, points up the risk in evolving possessive males. . . . Whenever a male is trapped and temporarily removed, his females are taken over by the males of his troop. After only a few hours the new possessor will not release the acquired females without a fight, even if the former leader is returned. The outcome of the fight will then determine whether the females remain with their original possessor or with the new one.

Field experiments among hamadryas baboons suggest that pair-bonds are perceived and respected by other males. Once these bonds can be observed in the behavior of the pair, males respond toward a pair or a bonded one-male unit not as separate individuals but as a single unit. (Lorenz calls this a form of *Gestaltwahrnehmung.*) This behavior appears to be a precursor of respect and serves as a stabilizing behavior that protects the pair-bond. Adult male hamadryas baboons thus appear to have evolved a mechanism for coping with the problem of highly possessive males: *Possessiveness is complemented by respect for possession.* An obvious inference from these findings is that the more general phenomenon of respect for possessions may have evolved from *respect for exclusive access between individuals linked by strong affiliative bonds.*[4]

It is seemingly a short step from hamadryas behavior to human possessiveness. Among humans, the concept of possession may have evolved from the "right" (in the eyes of the observer) of individuals to possess exclusively what they carry with them. The visual stimuli of perceiving individuals together, or an individual and an object (container, tool, weapon, or prey) as a gestalt, may trigger brain chemicals that lead to characteristic behaviors. That such behavior is often intense is illustrated by a report on chimpanzees. During a period of field observation of the chimpanzees in Gombe, field workers fed the chimpanzees considerable quantities of bananas. Some animals consistently tried to hold on to more and more bananas, which repeatedly slipped from their grasp. Although the animals appeared to be frustrated and almost bewildered by the situation, they persisted in their attempt to carry off and possess more than they could hold. The fact of "possession" also led to aggression. Goodall reports that the only time in 20 years of observation that she was seriously attacked by a chimpanzee was when she attempted to take a banana dropped by a female who held dozens in order to give one to a young chimpanzee who possessed none. The greedy chimpanzee can be said to have regarded the bananas she was carrying as her possession or property.[5]

The hamadryas baboon's propensity to behave in certain ways appears to be a consequence of genetic predispositions that are refined in predictable upbringing environments, resulting in a high degree of similarity in behavior among adult males. Not all species of baboons behave like hamadryas, however. Anubis baboons, for example, live in multimale groups, which is also a frequently encountered living arrangement among nonhuman primates.[6] Although there is pair-bond formation during female estrous, mating is random and, unlike hamadryas baboons, long-term pair-bonds are not observed.

In Chapter Two, the topic of modifiability of species-characteristic behaviors was addressed briefly. An excellent illustration of what is involved is provided in the following paraphrase of a report by Kummer. The report deals with two different species of baboons that came into contact in a new environment. They interbred, and this resulted in changes of their sleeping habits yet did not alter their family structures. Anubis groups, which roosted in the forest every night and made daily foraging trips into the thornbush, inhabited an area that consisted in the main of a gallery forest with trees up to 60 feet in height. Below a set of waterfalls, a canyon formation with steep slopes slowly deepened and widened toward a nearby desert. Hamadryas baboons who lived in the canyon spent the nights in cliffs. The major difference between the habitats above and below the falls lay in the proportion of trees and cliffs. Immediately below the falls was found an anubis group, however, that slept in the cliffs rather than trees (the anubis' normal and preferred sleeping place), although some suitable roosting trees were available. The social organization of the anubis gave no evidence of one-male units. Three nearby groups from the canyon were hamadryas-anubis hybrids. All the hybrid groups roosted in cliffs, as did the hamadryas. Their social organization revealed a curious mixture of anubis and hamadryas characteristics. There were one-male groups, most of which were small and unstable and, although some males herded as actively as pure hamadryas, their success in forming harems was meager.

Research into the behavior of a number of animal species suggests that environmental conditions, particularly those dealing with food supply and climate, influence within-species type of family structure[7] (although environmental changes generally occur slowly and selection for particular family types may interact with environmental change). For example, no anubis population in Ethiopia or in other parts of Africa has yet been reported to form stable one-male groups, nor have any hamadryas troops been found without one-male units. These findings suggest that both the anubis and the hamadryas baboons are capable of adapting their sleeping habits to local environmental conditions while at the same time maintaining a relatively inflexible social organization. Where genetic mixing between two different species (anubis and hamadryas) resulted in a few hybrid units, the social organization was not stable even in a stable environment. This situation may be analogous to human situations that indicate that economic factors were/are decisive in shaping family structures. The relationship between economic factors and family structure has long been apparent in traditional societies, and similar relationships may explain trends toward open marriage in modern Western societies where economic conditions either force women to support their children without male assistance—extreme poverty—or male family support is no longer necessary—affluence.

MOTHER-INFANT BONDS

As suggested earlier, a critical factor in understanding the family unit is the mother-infant bond. Of the many primate family structures that have developed in the course of evolution, one common factor has persisted: the long period of dependency of primate young on their mothers. Regardless of its structure or composition, the focus of the family as a unit is strongly rooted in procreation and care of the young until they reach reproductive age. Observations of nonhuman primates suggest that a mother and an infant (or any adult and infant) when together in close bodily contact, are responded to as a unit. This point is applicable across primate species. The perception of this unit is met with special behavior by other group members, and it often serves to inhibit aggression. Mothers in close bodily contact with infants are rarely attacked by other members of their group although at times other female group members will attempt to share infants with their mothers. This inhibition also works when young adults other than the mother hold infants, as is seen among low-ranking, young male baboons who occasionally snatch infants and hold them to avert aggression by stronger animals.[8]

BONDS AMONG FAMILY AND NON-RELATIVES

As central as the mother-infant bond is to the family, it is only one feature of most nonhuman primate kin-relationships. In typical families, juvenile sisters and brothers, as well as aunts, protect younger siblings from attacks or abuse by non-kin. Grandmothers hold, protect, and defend grand-offspring. And males protect not only their own offspring, but those of others. Literally all of these behaviors are predicted by kin-selection theory. Unidirectional investment should be preferentially directed toward kin, which is exactly what occurs. Unidirectional investment should flow predominantly from older family members to younger family members, which also is exactly what occurs. And, infants object to weaning and siblings engage in competitive behavior for the attention of parents, also behaviors predicted by kin-selection theory. Kin-bonds thus extend well beyond mother-infant bonds and provide a basis for group order, cross-animal obligations, and predictable interactions.

Bonds not explained by kin-selection theory are readily explained by reciprocal altruism theory. While reciprocal relationships undoubtedly occur among nonhuman primate kin, they are difficult to separate from kin-relationships. Thus they receive less attention than they deserve. Non-kin reciprocal relationships are quite observable, however. Pairs of animals become grooming partners. Animals that engage in aggressive behavior toward each other reconcile after fights. Animals that first identify a predator give warning calls to other members of their group. And, during periods of low food supply, all adult members of a group frequently reduce their food intake. Nonhuman primate groups thus are held together by a set of complex bonds, some based on genetic relatedness, others based on reciprocal relationships.

The picture of nonhuman primate behavior would be incomplete if the selfish side of animals' behavior were not addressed. High-status males and females have priority access to desired foods (e.g., fruit), to preferred sleeping sites, and to grooming by lower status animals. Animals will fight over access to other animals and over specific subterritories. Unfortunately, these features of their behavior were strongly emphasized in many early reports on nonhuman primate behavior. One consequence of this emphasis was that nonhuman primates were viewed as more selfish and exploitative than subsequent research has indicated.

The preceding review of nonhuman primate behavior can be given perspective in the following way. On one hand, there are a number of evolved tendencies that foster pair-bonding, group cohesiveness, and cooperation among group members. On the other hand, there is a part of nonhuman primate behavior that is selfish, exploitative, and largely self-serving. These tendencies are often in conflict and, at least insofar as nonhuman primates are concerned, evolution has not resulted in a predictable mechanism (or set of behavior capacities) that functions to reduce this tension. In subsequent chapters, the role of law will be viewed as providing such a mechanism.

Shifting the focus to humans, anthropologists generally agree that early humans lived in small groups, seldom exceeding 50. For such small groups to have existed it was necessary for humans to have evolved capacities—mechanisms—for assuring group cohesiveness. Many of these mechanisms are likely to be associated with mother-infant relationships, the pair-bond, and, ultimately, group structure. In a way analogous to anubis baboons, humans adjust to new homes or to environmental and social changes resulting from technology. Yet humans resist attempts to change certain fundamental relationships such as the mother-infant bond. Legal efforts to induce such changes have usually resulted in failure.[9] The bonds that assure cohesiveness and enhance survival are so strongly a part of our nature and so essential in the day-to-day conduct of human affairs that, at a very early age, individuals accept these bonds as an essential part of human nature. At times, of course, things go awry. Mothers desert infants, there are divorces, children run away from home, and kin murder one another. Nevertheless, the bonds remain and both direct and constrain behavior.

EXTRAPOLATIONS TO FAMILY LAW

How might the preceding points contribute to our understanding of the human family and laws that relate to the family? This question is addressed in greater detail in Part II. At this juncture, however, several key points may be noted:

First, regardless of environmental changes, the nonhuman primate family persists. The mother-infant bond is pervasive and this unit is by far the most stable among mammals. Where there is a pair-bond, even if it is only of short duration, the family unit usually comprises an adult male. These features evolved over long periods of time and they are the result of interactions between genetic predispositions, physiological events, and learning. And, in principle, they are similar across primate species.

Second, the forms and structures of family units differ. In nonhuman primates these differences exist in closely related species as well as among hybrid populations, such as hamadryas and anubis baboons. In humans the plasticity of bonding behavior is far greater than among other species. This results in a greater variety of forms of bonding compared to other species. Still, there are only a limited number of ways to bond. Environmental and cultural factors are important in determining which of the family structures found among humans prevails. Legal efforts to change existing family structures have been met with resistance and are generally unsuccessful.[10]

It is important to note that much of our understanding about the family and special bonds is assumed in the existing laws.[11] A mother's obligation to her children is hardly questioned, and laws focus on those situation in which mothers have not met their obligation. Similar points apply to a father's obligation to his young offspring. In other relationships, such as adult siblings, laws generally do not require that special obligations be honored. Nevertheless, there are strong social constraints against certain behaviors, such as rejecting a brother or sister in times of need. Persons who engage in such behavior will often find themselves socially ostracized.

Third, although evolution has developed some mechanisms that serve to protect both the pair-bond and the mother-infant bond, our genetic endowment alone is unlikely to be able to protect the human family. Yet most human mothers will protect their children. They often fight before they give them up. At times they will give their own lives to save them. So will most human fathers. In their struggle for scarce resources, they will tend to procure whatever is necessary for their own kin in preference to sharing with others. These propensities can lead to violence and, at times, even murder. It is in such arenas of potential and actual conflicting behavior that law often attempts to enter and to facilitate conflict resolution and coexistence.

Fourth, family structure is based on the functions of the different family members. There is a head of the family; other members assume various, sometimes submissive roles. In our time, human spouses may share the dominant role, each striving to maintain his or her supremacy in certain areas of decision making. Yet universally, family structure is the manifestation of social organization in a hierarchical or dominance order. This is what occurs when primates form any type of group (peer groups, clubs, athletic teams).

CONCLUSION

In this chapter I have discussed some of the behavior characteristics that we share with closely related primate species. Particular emphasis has been placed on bonds between individuals, their ubiquitousness among primates, and the importance of such bonds in human social interactions. Bonds, such as those between mothers and infants, are central to our understanding about the way our species behaves. In turn, our understanding has a direct impact on the law, both in terms of the laws we make and in our assessment of others' behavior. Much of the bonding behavior that was described can be accounted for by kin-selection theory and reciprocal altruism theory. Finally, a sketch of the evolutionary events leading to the development of capacities that enable us to make and to follow laws has begun to emerge. The role of law in this development begins to take shape.

NOTES

1. DeVore, I. 1972. Quest for the roots of society. *The marvels of animal behavior*. Washington DC: National Geographic Society.

2. Kummer, H. 1971. *Primate societies*. New York: Aldine Atherton.

3. Ibid.

4. Gruter, supra note 24, Chapter 1.

5. Goodall, personal communication.

6. Edey, supra note 44, Chapter 1.

7. Wickler, supra note 35, Chapter 1.

8. It has to be noted, however, that the behavioral disposition that protects the young from violence and possible death does not work in all situations. In various primate species and some other mammals, incidents were observed in which new alpha males who took over the harem of the predecessor killed the young animals (the offspring of the previous alpha) even if (or just because) they were in close bodily contact with their mothers. (These studies are sometimes used in support of kin-selection theory.)

9. Buxbaum, D. 1968. *Family law and customary law in Asia*. The Hague, The Netherlands: Martinus Nijhoff. Meijer, N. J. 1971. *Marriage law and policy in the Chinese People's Republic*. Hong Kong: Hong Kong University Press.

Massell, G. 1968. Law as an instrument of revolutionary change in a traditional milieu: The case of Soviet Central Asia. *Law & Society Review* II:2.

10. Ibid.

11. For example, the law has drawn a line distinguishing between conflicts within the family and among non-kin; for example, some contracts entered into by family members are at times not enforceable; also some cases of tort are not recognized if committed within the family. Family members cannot be forced to testify against each other.

FOUR

Legal Behavior, Group Order, and the Sense of Justice

The events of evolution that are of immediate concern, namely the evolution of the human capacity to make rules and laws and to follow them, are many and complex. In the preceding chapters some of these events and their implications have been discussed. This chapter begins with an overview of crucial steps in the evolution of human behavior. It then turns to the subject of legal behavior, group order, and the sense of justice.

EXPLANATIONS OF BEHAVIOR THAT IS SIMULTANEOUSLY SELF-INTERESTED AND ALTRUISTIC

A central theme of evolutionary theory is that those behaviors that are either a direct consequence of genetic information (e.g., reflexes, such as withdrawal from a painful stimulus) or that are an indirect consequence of genetic information (e.g., language), and that are associated with greater reproductive success, will appear with greater frequency in the next generation. Those behaviors that are indirect consequences require some type of environmental priming. Thus some of us speak English, others German, and still others Swahili. When priming conditions are present, indirect consequences occur with approximately the same regularity as do direct consequences. Thus nearly all children learn to speak and use their native tongue. Only in the absence of others using language do children grow up not using language.[1]

The fact that genes tied to reproductive success result in a greater number of the same genes in subsequent generations does not mean that the results of evolution are perfect or ideal. Far from it, in fact. Reproductive success references a competitive relationship. For example, the genes in family A may result in a tendency toward nonviolence while those in family B may result in a tendency toward violence. If members of family B proceed to eliminate most members of family A and also reproduce, more of family B's genes will appear in the next generation. From this perspective, evolution can be viewed as a competition between genes. It is also from this perspective that the two theories of altruism (kin-selection and reciprocal altruism) introduced earlier are so important in explaining human behavior in general and legal behavior in particular: From an evolutionary perspective they allow us to explain why such behaviors have evolved; from the legal perspective they allow us to explain behavior that is simultaneously altruistic and self-interested.

The fact that many behaviors that interest us are either direct or indirect consequences of genetic information does not mean that they are the same across all persons. All species, and especially *Homo sapiens*, are characterized by degrees of genetic, and hence behavioral, variance. Reproduction results in a constant mixing of genes. People differ in height, the capacity to sing on key, and to persevere. More important, they differ in capacities to make their own rules and to understand and to incorporate rules and laws into their behavior. This variance is an inevitable consequence of species plasticity due to the mixing of genes, and evolution that has increasingly favored the selection of genes that indirectly influence behavior.

Yet increased plasticity and behavior resulting from indirect genetic consequences do not imply that behaviors are ever fully free of genetic influence. Degree is always the issue. The nervous system of any animal species is a product of evolution, and behavior in all its forms and manifestations is a product of the activities of the nervous system. Our phylogenetic heritage is always a participant in our behavior.

We come then to the following situation. Basic self-interested behaviors have been strongly selected in our own as well as in all other species. The net effect of these evolutionary events is that as individuals we continue to be strongly predisposed to behave in self-interested ways, such as to assure our own and our kin's survival, to control resources, to engage in and win competitive interactions, and so on. Yet one outgrowth of these same evolutionary events is that we are capable of acting altruistically. It is these altruistic capacities that allow us to engage in behaviors different in degree from closely related species and to develop and follow many of the complex rules and laws characteristic of human societies.

From this perspective, our capacity to develop and follow rules results from two features of evolution: strong predispositions to act self-interestedly and equally strong—if not stronger—predispositions to act altruistically. This perspective allows us to place law in an evolutionary context. Law mediates our actions between those parts of us motivated by self-interest and those parts motivated by the need to act altruistically. Key to these events is our interdependency on others to achieve self-interested biological goals, such as acquiring a mate; reproduction; acquiring, holding, and dispensing resources; living in an optimally dense environment; developing non-kin social support networks; and investing in kin.

LAW AS THE MEDIATOR BETWEEN THESE TWO MOTIVATIONS IN BEHAVIOR

Law is thus a mediator. It is imperfect because it is created by the minds of members of an imperfect species and because individuals and groups often attempt to develop laws for reasons of self-interest. Laws are followed because the followers view them as providing a competitive advantage or at least no reduction in existing competitive status. There are links between such behavior and genes, more direct in some cases than others, but links nonetheless. It is in the context of the preceding points that I will now discuss legal behavior, group order, and the sense of justice.

LEGAL BEHAVIOR

The earliest humans appear to have lived in small groups, perhaps at times in groups as small as two individuals, composed only of a newborn infant and a mother. As groups increased in size, greater predictability of behavior of all group members became essential. The evolutionary event that was essential for the emergence of the human capacity to predict the behavior of others was that of developing the capacity to create and abide by rules, the impact of which extended far beyond the individual or the mother-infant bond. To be advantageous to our ancestors, such rules had to be understandable, shareable, and relevant to the requirements of social living and adaptation to the environment. A reasonable historical scenario of the setting in which human law evolved thus can be found, first in the dyadic relationship, later in small families, and still later in larger groups including non-kin.[2]

Among the earliest families, existence and survival were both shaped and constrained by mother-infant bonds. Then, as now, the newborn infant was dependent on the help and care of its mother. Human infants cannot cling to the mother, as is the case among many nonhuman primate species. It is several months before they can move about on their own. Mothers have to hold and transport them, to provide protection and nourishment. At times it was necessary for mothers to lay down their infants to perform tasks. Such tasks could not be performed indiscriminately. Early hominoid infants, like the infants of all primates, required monitoring. Family and larger groups that developed among early humans had to incorporate these behaviors into their social and survival strategies.

As I have stressed, a critical feature in the development of groups of increasing size and complexity was the necessity of predicting others' behaviors. Knowing how another person will respond to situations or actions facilitates both group cohesiveness and the possibility of achieving personal goals. For example, awareness that another person will react strongly and with great determination to defend the possession of property desired by both persons will affect the course of actions of participants. If a person desiring another's property can predict that the possessor will defend the property, he or she is likely either to desist from trying to take it or will attempt to acquire it at a more opportune time, using strategies such as deception.[3] Predictability of others' behavior thus is an important contributing factor to the formalization of rules: predictability is an essential requisite for laws to be effective.

At some point in hominoid evolution, cooperative foraging and hunting developed. Occasionally observed in chimpanzees, such cooperative searches and "harvest" of food became progressively more central in the transition from the australopithecines to our own tool using ancestors (*Homo habilis, Homo erectus*, and ultimately *Homo neanderthalensis*). Members of the groups not only foraged or hunted together, but they began sharing their catches according to implicit or explicit rules. Division of labor, cooperation, and planning became a necessity. Cooperation in hunting groups required an advanced form of communication. It is likely that there was a selection premium on individuals who were capable of learning quickly and who were capable of changing as the environment and the dynamics of social groups changed.

The events described above were only possible because of evolutionary changes in the brain. Neuronal wiring and structures had to become more complex, and neurochemicals had to serve an increasing array of functions. Increased cerebral specialization was not only essential but in part the basis for newly evolved capacities, namely to delay, imagine future events, calculate the potential costs and benefits of particular situations, recognize others, communicate the meaning and behavioral expectations of rules, incorporate laws and apply them to one's behavior, and many other activities.

THE DEVELOPMENT OF THE TRIUNE BRAIN

Among evolutionists, there are a number of theories that address the details of these evolutionary events. Each is somewhat controversial because we are unable to recreate exactly the details of our evolutionary past. The ideas of Paul MacLean are perhaps closest to the views being developed in this book. In MacLean's view, there has been a progressive layering of the cortex as hominoid evolution has moved from reptiles to mammals; the three key behavioral changes during this evolution were development of nursing, parental care, and play.[4]

A clear implication of MacLean's ideas is that much of our phylogenetic past is retained in the human brain, yet many new and very complex systems have evolved as additional parts in today's human brain. At times, the various parts of the brain act in concert. At other times, they are in conflict. And, at yet other times, the brain orchestrates destructive behavior. It is such variations that set constraints on human cooperation. The law inevitably has to address behavior resulting from such imbalances and conflicts.

As groups increased in size and as different members of groups divided their labor, hierarchical relationships developed along with the capacity to give and to obey commands. Groups whose members were inclined to obey rules and laws and who learned to control their actions within the prevailing rules of group life were more likely to survive than those who did not. Legal behavior thus offered a selective advantage.

INFORMATION TRANSMISSION, COOPERATION AND LEARNING

The preceding points can serve as the background for several topics that deserve special attention. These are: the capacity of the brain to transmit and encode information; cooperation and learning; language; and the concept of a sense of justice. These issues will be discussed in order.

The degree to which humans transmit information and use value judgments in accepting or rejecting traditional behavior is in large part a function of the neocortex, the youngest part of the human brain. It was the increase in size and complexity of this part of the brain that enabled our ancestors to increase dramatically their capacities to generate and use symbols and abstract thoughts. The brain developed to its present state over millions of years. Fossil discoveries, like those made by the Leakeys at Olduvai Gorge,[5] together with newer findings,[6] document the many stages of development.[7] These findings date back well over 1.5 million years. The discovery of tools and weapons associated with hominid fossils[8] suggest that our ancestors, who lived approximately 2 million years ago and whose brains were comparable to that of today's gorilla, were already engaged in a variety of novel behaviors compared to their immediate ancestors: "There is nothing in a quartz pebble or flint concretion that suggests the function of the knife, axe, or scripter that it may become. The fashioning of a useful tool from such a pebble can be the product only of both accidental observation and purposeful experiment."[9]

Findings like those cited immediately above imply not only that the development of human society was a slow process, but also that laws had a similar history. It is reasonable to assume that during the earliest moments of our legal heritage, a key factor was the transition from small groups composed of family members and perhaps a few non-kin, to groups in which not all members were known to each other. Two other factors seem essential to this development: predictability and increasingly accurate information transmission. An analogy will illustrate the importance of these developments: The reason that one is often frightened in new social settings is that one does not know how to predict the behavior of others. Frightening experiences are often turned to less frightening ones if strangers communicate by facial expressions or gestures their good intentions, for instance by making eye contact, offering a hand, smiling, and standing at an appropriate distance. This makes their behavior predictable. The development of capacities to transmit such information as well as correctly encode it are likely to have occurred in part because of a change in the brain's capacity to process more complex information, to make reasonably accurate estimates of others' motives, to assess the consequences of particular behaviors, and to incorporate past experiences into living and survival strategies.

Across these many evolutionary developments of the central nervous system, the general propensity to formalize rules as well as to develop and enact specific laws became part of our biological endowment. The rules with which people first complied are likely to have been of a specific type. They had to be of such a nature that the majority of members of a group could understand and follow them. They had to be rules that individuals followed because they believed that doing so would result in benefits.

Together with the growing ability to plan actions to meet with predictable responses, several other factors appear to have contributed to the early development of legal behavior. *Social cooperation* and *learning* can be singled out. As noted earlier, there is strong evidence that animals of many species engage in such cooperation. Cooperation[10] has both ultimate and proximate explanations. The two main ultimate cause explanations, kin-selection and reciprocal altruism theory, have been discussed. Cooperation is in part explained by kin-selection theory in which individuals invest in (help, assist) kin. Reciprocal altruism theory explains social cooperation among non-kin: A helps B at time X; at some subsequent time B repays A. For reciprocal behavior to be advantageous to the person offering help, the rules of such interactions need to be specific, and the social consequences to the receiver must be equally specific if the receiver does not repay the help received. The relatively ubiquitous nature of these types of behaviors, both across species and among humans across cultures, suggests ultimate causation (that is, selected behaviors to which we are strongly predisposed). Because the calculation and interpretive features of reciprocal altruism appear to be more complex than those associated with kin-selection, it is likely that reciprocal behavior evolved somewhat later in human history.

Observations of animals in their natural habitats suggest that one of the prerequisites of cultural behavior, the readiness to accept and utilize information developed by others, also exists among nonhuman animals. For example, the herd of elephant cows in which the young elephants grow up

> serves as a repository of traditional knowledge vital for survival: the routes to waterholes and feeding grounds, the seasonal movements to new ranges. It is possible to build and transmit such a repository because the elephant's brain [three times the size of man's brain] enables it to learn by imitation and to remember. Further, because elephants are potentially long-lived animals and mature slowly, several generations can exist in the same herd, increasing the young's opportunity for learning.[11]

Or, in another well-known instance, macaque monkeys adopted a new eating habit within six years after it was initiated by one young member of the group.

> Since 1952 an island troop of macaques has been provisioned with sweet potatoes . . . at a feeding station on the beach. One day Imo, a one and a half year old female, was seen washing the sand off a sweet potato. This new tradition was first imitated by one of Imo's playmates, then by Imo's mother. The potato-washing behavior continued to spread, generally from youngsters to mothers and from younger brothers and sisters to older ones. Four years later the only adult animals that had learned to wash potatoes were mothers of potato-washing youngsters. . . .

> When [youngsters] became mothers, they waded into the sea to wash potatoes with their young clinging tightly to them. The offspring imitated their mothers as if monkeys had always washed sweet potatoes.[12]

The macaques also eat wheat when it is scattered on the beach, where it becomes mixed with sand. The members of Imo's troop laboriously scoured the wet sand for single grains of wheat. Then Imo, at age 4, was seen to pick up a handful of sand and wheat, toss it into the sea, and scoop up the floating wheat after the sand had sunk. In the next few years most of the younger monkeys started "wheat-sifting." Adult males of the troop, however, including those of the highest rank in the dominance hierarchy, generally did not enter the sea, even after their favorite foods, a finding that directs our attention to the conservative side of animal nature: Animals bound by tradition often resist the development of new behavior patterns.[13]

The learning described above for elephants and nonhuman primates can serve as a framework of the type of learning characteristic of our distant ancestors as well as a basis for comparison of present-day hominoids with these ancestors. Many of the behaviors observed in elephants and nonhuman primates we see repeated innumerable times daily in human settings. Children imitate older siblings and parents. They learn the rules of the game or fair play when playing with their playmates.[14] We also see much more complex behavior. Adults take on new jobs that have new rules. People move across the country or countries and adjust to different sets of social rules. Integral to these changes is learning the rules of the social system one becomes part of, its requirements and rewards, as well as the punishments for not playing fair. It is in this area that one finds the proximate mechanisms for rule- and law-following behavior. The dos and don'ts of playmates, spouses, bosses, and the local police are all proximate in nature.

The development of language (which has both ultimate and proximate contributions) is yet another requirement for the development of legal behavior. Humankind's ancestors in the early stages of their evolution communicated with sounds, gestures, and postures. Through the refinement of these behaviors, language developed. With this development, rules took on different characteristics. Language could serve as the basis for formalizing rules and developing more complex codes; communicating important information before particular situations were experienced (e.g., storing food for winter); and developing laws to deal with as-yet-unexperienced situations. In effect, behavior, strongly influenced by behavioral propensities associated with kin-selection and non-kin reciprocation, as well as information dealing with survival in different environments, could be transmitted and specified both within the immediate context as well as for future events (e.g., handling hypothetical situations). Central to these events was the development of the capacity for language and the cognitive infrastructure that supports it. Language not only made it easier to make rules, to link them together, adapting them to various life-styles and environmental conditions, but also to enforce them (e.g., tell others that certain people disobey rules).

THE ROLE OF LAW

The ability to formulate, articulate, and understand rules, like other regularities that characterize successful species, undoubtedly had a significant impact on subsequent hominid evolution. The formulation of rules and, at a later stage of development, laws contributed to the development of social cultures rich in symbolic content. Language and symbol manipulation is everywhere apparent in these developments. An obvious outgrowth of such rules and laws is that they result in elaborate social traditions. Moreover, the adaptability of individuals and groups is influenced by these traditions. In effect, legal behavior is inherent in cultural behavior and law is inherent in culture. Law and culture interact. The interaction is dynamic. Yet it is not without conflict, a point underscored by a comment by Friedman:

> Americans . . . seem willing to pay their taxes; they evade, but within acceptable limits. . . . On the other hand, any attempt to use law to eliminate adultery in the United States creates entirely different problems of enforcement. No one obeys adultery laws simply because they are laws. . . . State intervention in private sexual behavior is culturally disapproved.[15]

Law is thus not simply a set of spoken, written, or formalized rules that people blindly follow. Rather, law represents the formalization of behavioral rules, about which a high percentage of people agree, that reflect behavioral propensities, and that offer a potential benefit to those who follow them. (When people do not recognize or believe in these potential benefits, laws are often disregarded or disobeyed, unless a ruling group employs strong means of enforcement, often supported by religious or ideological dogma.)

The capacity for language brought with it the capacity to imagine and to believe in the supernatural. This belief and the emotions associated with it were influential in shaping behavior, including legal behavior. Beliefs in the supernatural are closely related to feelings of worship, awe, and fear. A reasonable assumption is that the capacity to develop and follow laws was associated with unpleasant feelings (guilt) when laws were not followed. In turn, humans may have tried to escape guilt feelings resulting from those "tragic choices"[16] by allocating the responsibility for non-law-following behavior to a higher authority, the gods or the devil.[17] Perhaps humans first believed in evil deities, but in time, trying to live on good terms with them, they learned to believe in a just god. The emotional need to live on good terms with supernatural forces and to accept these immutable forces may be closely related to a willingness to respect authority and power. Imagination, therefore, likely served to embellish early laws with conviction and persuasive power.

Our understanding of rules cannot be divorced from issues relating to social structure, a point that applies to human and nonhuman primates. Early human groups undoubtedly were structured in dominance or hierarchical patterns in ways analogous to groups of nonhuman primates. Each individual held a certain place in such structures, which changed from time to time due to the aging of animals and new births. Those who were most dominant were likely to utilize existing rules to their advantage as well as to formulate new rules and enforce them. Although not all social rules among primates necessarily correlate with dominance, many do, and generally dominant animals are more likely to have their way than are subordinate animals.[18] Yet having one's way is not the only issue. For example, studies of macaques, baboons, and chimpanzees indicate that group-living primates rely on their leaders to find daily and seasonal routes to food sources. Troop-related decisions are most frequently made by individuals in high status positions.[19] Thus the position of high status is associated with obligations to lead and to carry certain responsibilities.

Rules and laws thus are in part reflective of social conditions as well as social options and expectations. If the law addresses itself to what is likely and unlikely in human behavior, it should take cognizance of these facts, and generally it does. For example, there are special legal rules regulating the behavior of minors, and the law demands specific responsibilities in the conduct of experts when they testify in court. Most licensing requirements, such as medical boards or bar examinations, are part of this special treatment by the law. In this way, the law allocates greater obligations to those who make special decisions affecting group members.

THE SENSE OF JUSTICE

We come then to the concept of justice and the sense of justice. One may describe the sense of justice as a person's feelings about the fairness of events and the emotional reaction to events perceived to be unjust. In what follows, I will describe how this sense might have evolved as one of the traits that has facilitated human adaptation. It is probable that as hominid evolution proceeded, mechanisms evolved that assessed whether events were matching expectations of an equitable outcome. This step in evolution is most apparent in our individual sense of justice—a dynamic cerebral mechanism that complements behavior, contributes to the guidance of individual behavior within the group, and that strongly influences a person's legal behavior (i.e., whether to obey the law, evade, ignore, or disobey it). Moreover, this capacity appears to be activated by sensory inputs, and (given the continuous change of human environments) must have included the capacity for change.

When and why do we have these feelings and emotions concerning fairness, justice, or injustice? The answer to the *when* part of the question partially lies in the fact that the laws humans have developed and follow are human in their origins. Our modern laws embody values and traditions created by humans, whether they are the indigenous laws of traditional societies or the norms contained in the sophisticated legal systems of technologically advanced nations. Individuals' legal behavior is not only guided by their ability and willingness to engage in such behavior, but what they actually do, whether they obey or break the law, is directed also by their own sense of justice and by their response to the concept of justice held by the group to which they belong. When members of a group make value judgments about acceptable behavior—even that behavior based solely on expediency or survival—these judgments often coalesce into rudimentary concepts of right and wrong, the basic concepts of any doctrine of justice. There are, of course, differences among persons no matter how closely knit members of a group may be. Each member of the group learns to adjust a personal yardstick and a personal sense of right and wrong. Age and sex in part contribute to these differences; at times, economic factors or past experience play an important role. The type of family structure in which a person grows up, the attachments and bonds developed during a person's early years, together with hierarchical order and nepotism and reciprocity are embodied in the patterns of behavior stored in that person's mind and that need to be matched in order to feel that justice has prevailed. A person's sense of justice appears to be the product of a combination of innate mechanisms that trigger feelings of justice or injustice when confronted with different situations. These emotions are present when events match or do not match expected patterns stored in the mind as, for example, when the balance or equilibrium in one's group is threatened or one has not been repaid for helping another. The desire for a balanced state of affairs motivates a person's actions on behalf of justice, and the restoration or continuation of this balance evokes feelings of well-being or gratification in the individual mind. In this context, pattern matching is the basic law.[20]

A possible answer to the *why* part of the question is found in recent studies designed to elucidate relationships between feeling states, specific brain anatomical structures, and physiological events. Findings dealing with a possible pleasure center located in the central nervous system and the functional effects of endorphins and enkephalins not only are of particular interest but also suggestive of a possible basis for positive feelings associated with the sense of justice. These findings show that an increase in the activity of certain endorphins and enkephalins is associated with feelings that are best described by such terms as well-being, confidence, balance, and pleasure. It is not farfetched to think that these and related physiological events have evolved and that the resulting positive feelings serve to inform individuals that events have occurred as they should. Such mechanisms should be present in all individuals. Their operation in any one individual will be influenced by factors such as age, sex, prior experience, and genetic makeup.

CONCLUSION

This overview of legal behavior, group order, and sense of justice has been developed to suggest that the behavioral response to law among humans is the result of evolution characterized by interaction among many factors, including genes, physiology, environment, and learned behavior. These factors coalesce in what we refer to as mind or brain and species-characteristic law-following behavior results.[21]

NOTES

1. Lieberman, supra note 4, Chapter 1.

2. See Chapter 5 for a detailed discussion of the structure and organization of the human family during this period of time.

3. Similar observations have been reported by de Waal and Goodall in studies of chimpanzees and other nonhuman primates. Although some interactions observed among chimpanzees can be interpreted as interference in conflicts by third parties, no definitive sanction for deviant behavior seems to exist. More clearly observed are the use of political strategies among the group to gain access to food, sex, and dominance.

Goodall, J. 1986. Social rejection, exclusion, and shunning among the Gombe chimpanzees. In *Ostracism*, eds. Gruter and Masters; de Waal, F. 1986. The brutal elimination of a rival among captive male chimpanzees. In *Ostracism*, eds. Gruter and Masters.

4. MacLean, P. D. 1983. A triangular brief on the evolution of brain and law. In *Law, Biology and Culture*, eds. Gruter and Bohannon.

> One might say in essence that the history of the evolution of mammals unfolds as the history of the evolution of the family. Taking care of the family, not only at the time of birth, but also as in the case of human beings, for a prolonged period, amounts to a progressive evolution of the sense of responsibility and what we call conscience....In the evolution of mammals vocalization and hearing became of utmost importance for maintaining parent-offspring relationships. Presumably vocal communication helped to insure contact among the diminutive early mammals living in the dark floor of the forest. The so-called "isolation call" is probably the most basic mammalian vocalization, serving to maintain maternal-offspring contact and group affiliation. . . . Most of the old cortex identified with early mammals is found in the great limbic lobe which is found as a common denominator in the brains of all mammals. In 1952, I suggested the term limbic system as a designation for this cortex and the structures of the brain stem with which it is primarily connected. (MacLean, 1952)

5. Leakey, L., and V. Goodall. 1969. *Unveiling man's origin*. Cambridge, MA: Schenkman. Potts, R. 1988. *Early hominid activities at Olduvai*. New York: Aldine de Gruyter.

6. Johanson, D., and M. Edey. 1981. *Lucy*. New York: Simon & Schuster.

7. Edey, supra note 44, Chapter 1.

8. "[F]ossil evidence is accumulating at an extremely rapid rate (more fossils were collected in 1970 than in all previous years put together)." Fuchs, L. H. 1972. *Family matters*. New York: Random House.

9. Hurlbut, C. 1968. *Minerals and man*. New York: Random House.

10. Axelrod, R. 1984. *The evolution of cooperation*. New York: Basic Books.

11. Eisenberg, J. F. 1972. The elephant: Life at the top. In *The marvels of animal behavior*.

12. Marler, P. R. 1972. The drive to survive. In *The marvels of animal behavior*.

13. Ibid.

14. Child development psychologists have researched this area dating back to the work of Jean Piaget in the early part of this century. In a chapter dealing with "The questions of a child of six," Piaget came to the conclusion that children ask why in order to find causal explanation, motivation and justification: "Let us designate as 'whys' of justification those which refer to some particular order, to the aim, not of some action, but of a rule . . . the child's curiosity . . . goes out systematically to all the rules that have to be respected—rules of language, of spelling, sometimes of politeness." Piaget, J. 1959. *The language and thought of the child*. London: Routledge and Kegan Paul.

15. Friedman, L. 1969. Legal culture and social developments. *Law and Society Review* 4:19.

16. Calabresi, G., and P. Bobbitt. 1978. *Tragic choices*. New York: Norton.

17. Calabresi, G. 1985. *Ideals, beliefs, attitudes and the law*. Syracuse, NY: Syracuse University Press. Calabresi introduces the "Evil Deity" who offers gifts, for example, the advances in technology, such as a car. Those who accept the gifts in good faith may become victims of unpleasant consequences, like being killed in a car accident; but so will others who don't accept the gift.

18. McGuire, M. T., M. J. Raleigh, and C. Johnson. 1983. Social dominance in adult Vervet monkeys I: General considerations. *Soc. Sci. Information* 22:89-121.

19. Kummer, supra note 2, Chapter 3.

20. Hoebel, B. G. 1983. The neural and chemical basis of reward. In *Law, biology and culture*, eds. Gruter and Bohannon.

21. Gazzaniga, M. *The social brain*. Supra note 2, Chapter 1.

PART II

Human Behavior and Legal Norms in Family Law and Environmental Law

Part I focused on biology, mind, and law. Part II will illustrate the application of ideas developed in Part I in two areas: family law and environmental law. These are not the only areas in which the relevance of ethology for law can be demonstrated. Ethology can enlighten law in numerous other areas, including criminal, contract, and property law. Family law and the functions of sexuality have been selected as one topic because family relationships are early as well as crucial components in the development of all social and legal concepts. Environmental law has been selected because of the many novel issues it poses for law. Chapter Five deals with family law and sexuality; Chapter Six deals with environmental law.

FIVE

Family Law and the Functions of Sexuality

Sexuality has four basic biological functions: pair-bonding; the exchange of genetic material; procreation; and care for offspring (broodtending). These functions represent important elements of social behavior in literally all mammalian species, and among humans the majority of cultures have implicit or explicit rules for each of these functions. In this chapter, the preceding functions are discussed more or less in the order above with the first part of this chapter addressing functions of sexuality and their influence on family structure and family law.

For the most part, laws dealing with sexuality are in the province of family law. (Criminal law is involved in situations in which a society considers behavior to be unnatural or perverse, or where violent acts are committed, as in rape.) Family law has its roots in biology, yet biology is not the only factor influencing family law. Cultural influences are also apparent. Obvious biological facts are interpreted differently across cultures and given meaning in terms of prevailing philosophical, religious, and ideological percepts, and they are modified in response to practical options. Polygyny, for example, is part of

Islamic law and is practiced widely among some 800 million Moslems. Western jurisprudence sees bigamy as a "crime," even when practiced by those whose religious beliefs include polygyny (e.g., Mormons). At times, bigamy has been punished by death, as in James I's civil courts in England, and the State of Virginia in 1788.[1] Degrees of enforcement vary, in part because people disregard laws dealing with sexuality more than any other type of law. Nevertheless, laws dealing with sexuality and sexual behavior are present and they have a significant impact on the lives of all persons to whom they are addressed.

PAIR-BONDING AND MARRIAGE

In most human societies, pair-bonds are formalized in the institution of marriage and, in most cultures, marriage requires some type of legal sanction. These sanctions deal primarily with different consequences of sexuality, not with sexual behavior as such—if the primary function of marriage was to control sex, then marriage (but not sex) might have been abandoned long ago. Among preliterate human societies, pair-bonding is frequently based on the exchange of women between kin groups. Strong ties or bonds between the partners often develop during such marriages. Formal exchange also characterizes marriages in modern societies, as in Moslem countries, where high percentages of arranged marriages are customary across all social strata. (Just prior to her election as Prime Minister of Pakistan, Benazir Butto entered into such a marriage.) In industrialized societies, the degree to which the exchange of women (or chattel for women) remains a part of marriage is probably understated, although the methods used for exchange are less overt, for example, early influence on offspring regarding marital roles and partner choice, familial approval or disapproval of the future spouse, and the influence of peers often combine with motivations based on economic considerations.

Pair-bonding is found throughout the animal kingdom. The term will serve as an analogue for human male-female sexual partnerships or marriage. Traditionally, the law has addressed pair-bonding behavior mainly in the context of marriage. This context may be formal or informal. The formal aspect is familiar in the state-sanctioned institution of marriage. The informal aspect is seen among unmarried couples who, although refusing to legalize their bonds, often claim the same protection by the law as that given to those who are married. In the past, courts tended to deny rights that protect state-sanctioned marriages to persons who were not married (although the legal concept of common-law marriage has existed in the United States since the mid 1800s). Both types of sexual partnerships can involve pair-bonds. That this realization enters into legal decision making is suggested by the increasing number of instances in which alimony (and child custody) among unmarried partners has received legal sanction.[2] Legal decisions thus appear to reflect changing sexual mores and living arrangements among an increasing number of persons who are sexual partners.

Although legal norms applicable to pair-bonding have varied, throughout recorded history social groups have seen reproduction as the natural outcome of such bonding. Prehistoric art supports this interpretation through its emphasis on fertility, particularly in representations of the female body. Western society has placed and continues to place a strong emphasis on the function of reproduction in marriage. New factors, however, are changing expectations and mores. Recently in our history, great emphasis has been placed on a type of sexual attraction embodied in the word "romance" and its supposed importance for marital happiness. This emphasis, to the exclusion of other factors, has had predictable consequences. Primary among these is the disregard of pair-bonding as a crucial element in the creation of new attachments among persons and families, as a basis of mutual protection, as a means for the transmission of genes and for the care of offspring, and as a necessary condition for establishing lasting and altruistic relationships. Current Western views often overlook the fact that sexuality serves functions other than either romance or reproduction: for example, it can strengthen pair-bonds, and contribute to family stability and to the maturation of family members.

Bonds between male and female partners are seen throughout the animal kingdom and represent remarkable biological processes, usually characterized by "ceremonies" that each individual performs only with its mate or with some other closely related member of the family.[3] The duet songs typical of monogamous birds (whose sex cannot be distinguished externally), are examples. In some species, both partners sing the same notes or phrases, either echoing each other or singing in unison. In other species, the partners utter different phrases or parts of phrases, combining them in various ways, as is seen among monogamous gibbons.[4] In yet other species partners behave in similar fashion, as in the dance of adders.[5]

While the term *pair-bond* describes the joining of a male and a female, such bonds are not limited to adulthood. Members of some species engage in characteristic pair-bonding behavior before they reach sexual maturity, as is the case with adolescent hamadryas baboon males who attempt to acquire female mates before they are sexually mature. At times, these bonds may be lasting and respected by much stronger adult hamadryas. Similar behavior is seen in sexually immature greylag geese.[6]

The behaviors facilitating pair-bonding are multiple, complex, and generally triggered by hormone changes.[7] The details of such behavior are influenced by the behavior of others and by environmental variables (e.g., season, available nutrients), and they are modulated by maturational stages. In many situations specific contributing behavioral elements cannot be isolated, and individual behavior patterns often appear to be at cross-purposes (e.g., threats combined with courtship postures). Chimpanzees, for example, use essentially the same behavioral display in sexual rituals, intermale aggression, in periods of frustration, and in confronting natural objects.[8] Such similarities are unlikely to represent a limited behavior repertoire. Rather, like humans who hug others in various situations (e.g., when they greet old friends, or make love, or when they fight), chimpanzees also communicate the meaning of their behavior prior to its occurrence.

CONVERTIBLE BEHAVIOR

Wickler has noted that the function of some sexual behaviors has evolved to serve more than a single function, or may have unintended consequences. Lorenz has made similar observations concerning the behavior of greylag geese.[9] Behavior that evolved in broodtending (including human child rearing) or that stimulated sexual practices, can also serve to unite pairs over a longer period of time. As noted above, the display behavior of chimpanzees is used in several contexts. In her study of wild chimpanzees, Goodall noted that male chimpanzees practice such behavior at an early age and that until adolescence it is difficult to distinguish between play and aggression. During puberty, the frequency of display behavior increases. There are always aggressive elements in this behavior. Chimps wave their arms, shake branches, and throw stones. Such displays generally do not result in fights, however, and severe injuries have rarely been observed following such interactions. The usual response by the recipient is a threat. Display behavior thus appears to be one way in which animals express excitement, release tension, and communicate frustration and anger. At times it leads to redirected aggression, such as attacking objects or other (usually weaker) animals. (The behavior implies the presence of constraining mechanisms that serve to reduce the frequency of aggression on the part of animals who are capable of inflicting severe injuries on other animals.) The same behaviors also have a place in initiating courtship and sexual relationships. The male chimpanzee will shake branches and assume a threatening posture toward estrous females, his hair and penis erect. At times, he will even attack a female if his threat does not suffice to bring the expected submissive response. These behaviors are integral parts of courtship and the politics of negotiating access to estrous females.[10]

Threatening display behavior is also used in other contexts, for instance in efforts to gain access to or maintain a position in a dominance hierarchy. For example, at Gombe Stream a chimpanzee named Mike kicked and rolled empty cans as a display, thereby creating noise and confusion among his competitors. The confusions allowed him to climb from a low-status position to that of the alpha male, which he retained for 5 years.[11]

One of the least recognized facts of male nonhuman primate dominance hierarchies is that dominant animals fight with other adult males less often than subordinate animals fight with each other. Subordinate animals frequently make mild threats toward dominant animals. A threat or display in return is usually sufficient to terminate the interaction and to affirm the status of respective animals. Moreover, dominance relationships are comparatively stable. Such stability allows for predictability in behavior responses among group members. It also appears to have a pacifying effect (in the sense that, in day-to-day interactions, predictability generally is more desirable than unpredictability). Clearly, also, animals are willing to forgo considerable potential personal gain (e.g., changing from a subordinate to a dominant status) to participate in social relationships. Yet, perhaps the most important point about stable hierarchies and the associated predictable behavior is that, together, they serve to ameliorate the tensions that result among animals that are interdependent, engage in affiliative relationships, yet also engage in competitive, often aggressive, interactions. Both self-interest and interdependence are features of group behavior and, generally, the more complex a group's social structure, the more frequent are convertible behaviors.

Convertible behaviors thus explain a large number of seemingly paradoxical behaviors that are characteristic of family situations. There is pair-bonding, yet there is competition. There are threats and counterthreats, yet there is holding and hugging. There are interdependencies and fights over resources, yet there is sharing. Sibling rivalry and teenage challenges of parents are common. Still, there is respect and helping. There are periods when family members join together to defend the family against external challenges. And, on rare occasions, family members murder each other. Through this seeming chaos of relationships, bonds, and ministruggles, families remain families and generally preserve their ties.

BONDING AND THE ACTION CHAIN

Among chimpanzees, once courtship displays have begun, a sequence of reciprocal interactions follows. The claims of one partner become the obligations of the other. This forms what Hall[12] has called an *action chain,* by which each partner, responding to the actions of the other, participates in the progressive development of predictable behavior patterns. Available evidence points to the conclusion that it is adaptive for a species to have the capacity to use specific behaviors for different purposes and for individuals to know that a certain action will evoke a predictable response. Predictability not only facilitates reciprocal relationships among group members, but it also serves to create a balance among participants with justifiable expectations. Among humans, expectations are often expressed as claims or rights, at times as obligations imposed on others. When ceremonies and rites accompany reciprocal actions, they evoke the attention of others and may be accepted as mores or norms signifying that each partner is expected to engage in specific behaviors given certain conditions.

As noted, among chimpanzees the female responds to the male's courtship behavior by adopting a submissive, crouching position and permitting the male to mount her. In other contexts, a part of this sequence is used as a freely convertible behavior when female chimpanzees adopt a crouching posture as a submissive gesture during aggressive encounters, thereby signaling the acceptance of another animal's dominance and, in most instances, avoiding confrontation. Similar points apply to males: When a threatened chimpanzee demonstrates his intentions by adopting a submissive posture, he expects a specific reaction from the other, namely, an end to the aggression.

This expectation becomes the obligation of the aggressor who, in most instances, refrains from further attacks during the specific episode. Often a further specific action is expected from the threatening aggressor, such as a reassurance gesture. The dominant chimpanzee touches the head or the shoulders (or other parts of the body) of the less-dominant animal in order to reassure him, seemingly as a symbolic signal for the end of the confrontation.

PAIR-BONDING IN HUMAN SOCIETY

Human sexual behavior includes specific components that do not have direct sexual functions in the sense of sexual activity or reproduction, but which, nevertheless, are features of human sexuality. These include aspects of pair-bonding behavior, such as child rearing and care of the offspring until they reach social maturity. Much of this behavior appears to be a consequence of the continuous sexual receptivity of the female. In all nonhuman primates, females have brief yet clearly limited periods of estrous. Because this state is so universal, it is likely that this condition was present among our ancestors.[13] Cross-species studies suggest that sexual activity alone does not provide a sufficient reason for pair-bonding. Pair-bonds are not limited to the period of female ovulation (estrous) among hamadryas baboons. These bonds sometimes exist before sexual maturation and continue into adulthood. It is the exception, not the rule, however, that a given pair-bond extends over several breeding seasons. Perennial sexual availability of the human female thus appears to be a contributor to lasting, close personal relationships between human sexual partners.[14]

Regardless of which mechanisms contribute to pair formation and pair-bonding among species, the fact that the bonds exist in numerous species suggests that the pair-bond has valuable functions apart from reproduction. This fact may explain why some laws dealing with sexual behavior are less effective than others. For instance, deviance is high for those laws that emphasize only the reproductive function of sexual behavior and disregard or forbid the pair-bonding functions of sexuality. An example is found in the high number of illegal abortions that occur in countries that prohibit the interruption of pregnancies and in which women cannot have abortions performed under hygienic conditions.[15] Violations of antiabortion laws, especially when birth control devices are limited or unavailable, often are the result of sexual behavior that serves the function of pair-bonding or pair formation and not primarily that of procreation. Put in motivational terms: The motivation for sex (and the pleasure and tension relief often associated with it), need not be limited to the desire to procreate.

In Western society and in most other complex societies, the pair-bond and the different functions of sexual behavior in humans have been regulated for centuries by marital law. Ceremonial marriage has held an important place in all social groups, and anthropological studies of nontechnological societies suggest that the marriage ceremony has a long history. On an evolutionary time scale, this history is short, however. Shorter still is the role of the church in legalizing marriage. The sanction of the state as a requirement for secular marriage comes even later. In the progressive secularization of ceremonies and symbolism used in uniting partners in marriage, a clear trend toward public control and scrutiny is detectable.

How do the points above translate into our current laws? Today, not only marriage but also many other elements of pair-bonding behavior are subject to legal regulations. The state-approved wedding and the resulting financial obligations are but two examples. In most instances, the rights and obligations of the marriage partners are defined, as is their relationship with existing or potential offspring.[16] Current marital laws, or the law of domestic relations (as it is called in California), deal primarily with marriage breakdown and divorce—dissolution of pair-bonds—and the social consequences that often arise. Generally, they do not deal to a great degree with behavior of husbands and wives. Rather, the laws are primarily concerned with those behaviors that are basic to social interactions, and reciprocal protection essential for child rearing and the care of children and juveniles until they reach social maturity. Regulatory laws (e.g., the legal requirement for a marriage license) are not primarily concerned with behaviors regulated (at least in part) by a biological program. Many laws, however, that involve pair formation, child rearing, social interaction, and mutual protection attempt to channel behavior that is influenced by biological processes.

Other examples of laws that have regulated or continue to regulate sexual behavior are those outlawing premarital sexual relationships, adultery, incest, homosexual acts, rape, prostitution, and sexual offenses against minors. Here, criminal law enters the sexual sphere. Nowhere is legislation and adjudication as difficult or as controversial. Thus, it is not unexpected that violations of these laws are incompletely reported and that such laws are not uniformily enforced.[17] Extensive research will be required before it is possible to suggest approaches to reconciling the conflicting views in these areas. Generally, it can be said that any behavior-channeling laws that address human sexuality are effective only insofar as their dictates are adapted in a coherent way to the behavior they are intended to regulate. Obviously, legal efforts in stabilizing pair formation have not proven completely effective, because divorce, either formal or informal, is practically universal. Factors contributing to marriage breakdown, divorce, and its consequences, are the areas that require new studies.

What can ethological studies contribute to legal views on marriage and marriage breakdown? Studies are likely to further specify:

- the interaction of bonding functions with other biological functions;
- the importance of behavior leading to pair-bonding with respect to facilitating within-group interactions and in linking members of kinship groups;
- general stabilizing effects of predictable behavior chains; and,
- the roots of cooperation, division of labor, and altruism.

It seems that laws mindful of these aspects of pair-bonding will be accepted by the vast majority of people and thus prove relatively effective. Before these possibilities are discussed in greater detail, however, other biological functions of sexuality need to be examined.

EXCHANGE OF GENETIC MATERIALS AND THE LEGAL PROSCRIPTION OF INCEST

One of the points emphasized earlier dealt with cultural constraints and social support of pair-bonding. In part, these functions are mediated through laws dealing with the exchange of material property and, in part, through laws dealing with the exchange of genetic material. This section focuses on the background to laws regulating the exchange of genetic material.

Until recently, the transfer of genetic material among humans was solely a function of sexuality and the role of rules and laws was reflected primarily in the almost universal proscription of incest. A more recent phenomenon is the involvement of the law in situations created by new technologies, such as the various forms of artificial fertilization. New technologies have created numerous conflicts among advocates of different ideologies and religions.[18] Changes are reflected in a new vocabulary dealing with the exchange of genetic materials: biological parent, genetic parent, social parent, birthmother, adoptive parent, sperm donor, and sperm seller. New terms also apply to offspring rights, which will be discussed presently. The impact on genetic material transfer, which is an integral part of this development, is only beginning to receive the attention of lawyers and legislators.

The exchange of genetic material is one of the cornerstones of evolution. Exchanges alter genetic makeup across generations and, in turn, both the variety and variability of individual organisms within a species. In literally all known instances, the role of sexuality is a key factor in this exchange—even many of the oldest known living organisms (single-cell organisms), which reproduce by cell division and not by sexual means, at times merge and exchange genetic material.

Within limits, exchanges of genetic material increase the vitality of organisms and enhance their ability to adjust to different features of the environment. This point deserves some elaboration because of its implications for understanding behavior and the degree to which laws are likely to be followed. Species generally segregate into specialists and generalists. Specialist species are those that survive in specific ecological niches. Generalist species can survive in a broader range of niches. If the niche conditions are optimal, specialists multiply rapidly. If the niche conditions change, however, survival rates may be drastically reduced, at times even to the point of extinction (e.g., panda bears). Generalists, on the other hand, are capable of surviving in situations of greater niche variability. These differences in survival probability occur largely because generalists have a greater degree of genetic variability, which increases their adaptive capacities, compared to specialists. This variability increases the probability that a percentage of generalists will survive as niches change.

The preceding is but part of the story, however. Increased genetic variability is usually associated with increased plasticity. This point holds particularly for *Homo sapiens*. Increased plasticity is associated with increased cross-person differences in behavioral characteristics; for example, types and intensities of motivations and capacities for altruistic behavior. The plasticity of human behavior is reflected in the structure of Western legal systems. Legislation attempts to limit behavioral deviance by establishing specific goals of behavior for all citizens. Adjudication attempts to apply these laws but considers also individual motivations and characteristics (e.g., extenuating circumstances, sanity). The more specific or constraining a law is when dealing with strongly predisposed behavior, the more likely it is that individuals will devise ways of avoiding the law, and the courts will get involved in deciding whether the behavior is legal or illegal. This applies to civil law as well as criminal law.

Returning to the cause of behavioral plasticity—the exchange of genetic material—there is one very important feature in this process that has been addressed by law, seemingly independently from, yet in complete harmony with, the evolutionary developments—namely the issue of inbreeding. Organisms have developed mechanisms that reduce inbreeding—the transfer of similar genetic material. (Even plants are equipped with complicated devices to prevent self-fertilization.) These mechanisms serve to constrain the frequency of incest.[19] The genetic consequences of inbreeding among humans are well known, namely, an increased probability of birth defects and reduced adaptability. Such consequences offset selection favoring adaptive behavior. Some of the incest inhibiting mechanisms are inherent in the way parent-child bonds ordinarily dissolve. The young are weaned and this event increases the chances that offspring will have ambivalent attitudes towards their parents and thus reduces the probability of inbreeding. There are also strong cultural prohibitions against such behavior. Available data suggest that the inhibition against incest is neither totally rigid nor of equal force at all ages, and the different types of incest (father-daughter, mother-son, and brother-sister) may be regulated by different mechanisms that mature somewhat independently. Sexual play among children is a common phenomenon and it is probably part of the human desire to explore the environment. The case of brothers and sisters provides an instructive example. Certainly there is sexual exploration among siblings, and inhibitions may have to wait till puberty. Sorenson's[20] research, which was published in 1973, reveals that in a group of adolescents between 13 and 15 years of age, 69% of the boys and 84% of the girls expressed the feeling that sex between a brother and a sister was abnormal and unnatural. Among 16- to 19-year-olds, 86% of the boys and 85% of the girls expressed this attitude.

Allowing that Sorenson's data describe the attitudes of teenagers in a reasonably accurate way, one may assume from the attitude of the older group that behavior mechanisms inhibiting incest are not innate, rigid controls. Rather, they have a phylogenetic origin on which ontogenetic development builds during maturation. Normally, this development leads to behavior that coincides with the content of laws against incest. This may be the reason why compliance generally is high. Nevertheless, incest prohibition is far from foolproof and, where family ties are strong, the possibility of inbreeding remains.[21]

Lévi-Strauss[22] in speaking of the problem of incest states that the prohibition of incest, because its format characteristic is universal, has been taken from nature: Incest reduces genetic variability. Fox[23] concludes that the incest taboo has a biological basis and is now part of our cultural heritage: "It originated either because it was of selective advantage in preventing disastrous results of inbreeding, or because it was the inevitable outcome of demographic limitations of inbreeding." Although these two interpretations of the evolutionary events differ somewhat, the central point is that the results of both explanations are the same. More important, the fact that instances of incest occur does not preclude the probability of a biologically programmed mechanism for the ontogenetic development of an inhibition against incest. Some ethologists, for example, feel that humans develop strong sexual, psychological prohibitions during puberty and that these reflect the maturation of innate inhibitions. This opinion is supported by findings of ethnologists and anthropologists who have observed incest taboos in literally all the groups they have studied.

Although norms for many aspects of sexual behavior differ across human groups, the proscription of incest is regulated in almost all cultures by essentially similar legal norms. It exists in virtually all societies, from the taboo of hunter-gatherer groups to highly developed Western societies, where incest is a crime and subject to prosecution by the state. Both the Civil Code and Penal Code of the State of California, for example, define the degrees of consanguinity within which marriages between close relatives can be declared to be void or illegitimate.[24] Until 1976, acts of fornication or adultery between these persons were punishable by imprisonment of up to 50 years.

Prohibitions against incest have never been fully successful and, as mentioned, cases of incest occur in virtually all societies. In rare instances during recorded history, marriage between brother and sister was even prescribed for the ruling class or in religious cults. Dynastic incest existed among the Ptolemies, who inherited the empire of Alexander the Great and who frequently married their own full sisters. In ancient Egypt, kings sometimes married half-sisters. Cases of dynastic incest have also been reported in Peru during the Inca Empire and in Hawaii.[25] Nevertheless, instances remain rare and are interpreted as responses to special conditions.[26]

The preceding points are reflected in today's legal efforts directed against incest. In most instances, these efforts appear to be in harmony with the biological postulates serving the genetic material exchange function of sexuality and thus complement the biological functions of sexual behavior to which these particular legal rules are directed. The rules serve to advance the diversified exchange of genetic material, even though only recently—in evolutionary terms—have we become aware of the biological details of genetic exchange—scientists did not clearly understand fertilization or the rudimentary facts of genetics until the late nineteenth century.[27]

The counterpart to constraints on inbreeding, namely probable tendencies limiting degree of outbreeding, deserves mention. Just as there appear to be evolved constraints on inbreeding, similar constraints appear to apply to outbreeding that, when extreme, serves to dilute traits. The literature on kin-selection theory, which addresses this issue, implies that optimum breeding partners are something like second cousins, who would have a small percentage of replicate genes. Second-cousin breeding would facilitate both the preservation of certain desirable traits as well as the introduction of new, yet not extremely novel genetic material. (An elaboration of this point leads directly into often encountered cultural prohibitions against interracial marriages.) Many of the customs dealing with marriage, such as knowing the person, the person's family, or both prior to marriage thus appear to be behavioral manifestations of genetic-based tendencies to avoid diluting existing traits too severely. Although customs and mores have guided these manifestations among humans, law generally has not stepped into this arena. Only very recently in the evolutionary history of hominids has an extreme mixture of the gene pool become possible. Until a few hundred years ago, prior to the advent of modern transportation, most young adults found their partners within a limited territorial range.

Granting that diversification (within limits) of a species' gene pool is one consequence of reduced inbreeding, it is worth asking why rules against inbreeding have existed for thousands of years among humans, that is for a period that significantly predates scientific explanation. A likely possibility is that the human mind,[28] while developing rules and formulating laws to regulate behavior, often acts in harmony with biology. Only those rules and laws that are adaptive, and the societies that adopt them, will survive over long periods of time. Those cultures that developed laws against incest are likely to have gained a competitive advantage because they were in harmony with important biological functions. Translated into the context of this book, this means that the more we understand about the biological functions of human behavior the less likely we are to use law to our own destruction and the more we can substitute scientific knowledge for the chance factor in the development of law.

The preceding conclusion may be put another way. Humans can conceive of—and have adopted—an extraordinary variety of norms, some of which may seem to us absurd or ineffective. Customs and laws with regard to sexuality and the appropriate number of offspring, marriage, division of property, or rules for child rearing vary enormously across time and place. Yet, it is dangerous to presume that our perception of the world is the only natural one. Some cultural rules and laws may well be mistakes that entail long-term costs for those who adopt them. In this sense, one can think of nature as occasionally vetoing a law that has proved to be out of harmony with human behavioral tendencies.

PROCREATION

Although the details of the exchange of genetic material were not known to scientists until recently, most cultures have known the role that sex plays in procreation.[29] Moreover, the realization that sexual intercourse and pregnancy are related has been an underlying theme in the moral and ethical values directing Western marriage laws for over 2,000 years. During these years, biology, custom, tradition, and religious attitudes came together in these laws. In the first part of this section, these topics are discussed. In the second part, the discussion focuses on the tremendous changes in the social role of women during this century, partly due to the development of new contraceptives and partly due to changes in the male or father role.

The biology of procreation can be addressed at two levels. One focuses on sexual differences in procreation strategies, the other on physiological changes associated with pregnancy and lactation and the development of offspring. Sexual strategies will be considered first.

One of the fascinating outgrowths of modern evolutionary biology has been the elucidation of sexual strategies. The description of these strategies that follows does not imply that women or men intentionally pursue such strategies. The theories deal with explanations and genetic implications of behavior. According to this theory, a female may be expected to develop and act on strategies that will result in her attempting to share her genes (procreate) with the seemingly best available male genes. In following this strategy, she may find that the seemingly best available genes may or may not belong to her social mate. That this actually happens—namely, some women marry one person but have their children by others—has been established repeatedly, and on average 5% to 7% of offspring are believed to have biological fathers who are not their social fathers, and the social fathers are unaware of the fact.[30] The possibility that being a social and biological father may not be the same undoubtedly did not pass the notice of our ancestors (e.g., in Roman law this uncertainty of fatherhood was recognized: *pater semper incertus*). Moreover, such knowledge is likely to be one of the main contributors to male sexual possessiveness and jealousy: From a male's perspective, it makes little sense to be the social father but unknowingly not the biological father. This concern is reflected in early statutes of Roman law, where *ubi matrimonium ibi pater* as ruling doctrine provided stability for marriages and security for offspring.

To talk about female sexual strategies is not to assume that females choose extramarital sex partners with the conscious intention of having children. There certainly can be other reasons for this behavior, for example, exchanging sexual favors for material advantages and sexual attraction. Yet, the end result of such behavior may be better genes compared to those available in a social partner. Genetic material is not the whole story, of course. Once genes have been shared and a child is born, resource predictability will increase in importance. Thus a woman might wish the father of her growing child to be a person who provides resource stability and such a person may not always be the previously selected genetic father of her child.

Males, on the other hand, may adopt a different strategy (again, not necessarily as a conscious decision): In principle, the male strategy would result in males attempting to father as many children as possible by more than one mate, hoping that some of their offspring will grow to maturity and, in turn, reproduce. The female, as well as the male, response to these opposing strategies should be one of vigilance, possessiveness, and sexual jealousy; the male motivated by his unwillingness to support offspring fathered by others, the female motivated by her need to find both seemingly good genes and support for her offspring.

We thus have a species with different sex-related strategies. Within-sex variation is to be expected; for example, some males may desire and seek monogamy. Still, in the final outcome, procreation is a matched investment in genes by the male and the female parent. Laws dealing with procreation seemingly have at times reflected sex-related differences in the separate treatment of the sexes who engage in extra-marital relationships. Yet, on the whole, the law has played an important role in balancing conflicting interests that, according to sexual strategy theory, are part of our biological program. The creation of the legal concept of marriage and the numerous laws dealing with the care for offspring and responsibilities to such offspring have contributed to the stability of the family and, on balance, protected both parents and offspring.

Since early times, religions have built much of their power around the moving forces of human evolution. Once the state entered the arena of governing human behavior, effective secular law followed most of the existing behavior. When the state assumed legislative control of marriage and divorce, prevailing religious values generally were integrated into laws for ideological or political reasons, that is, the state adopted values similar to those of the church. In the West, for example, most of the legal norms that regulate sexual behavior reflect the traditional Judeo-Christian ethical premise that the major purpose of sex and marriage is reproduction.

In the regulation of genetic exchange discussed above, customs, mores, and laws reflect biological factors. Although the understanding and incorporation of biological information has differed across cultures, procreation as a function of sexuality and as a goal of culture has progressively gained an important place cross-culturally in laws regulating sexuality. Through monitoring and sanctioning such functions, our ancestors increased their capacities to protect their investment in offspring. From one perspective, that of population growth, it may be that the combination of maintaining and sanctioning reproduction may have worked too well. Yet the church often went further than secular law. The case of the Roman Catholic church is instructive. The prevailing dogma concerning all sexual behavior places the emphasis on procreation largely to the exclusion of other functions of sexuality. Those who imposed the rules no doubt realized that laws regulating sexual practices would be disobeyed. The imposition appears not to have been aimed merely at compliance by the majority, but rather at establishing a moral code that could not be flouted openly. Inevitably, the code would be violated by many, thereby making the dogma of human sinfulness evident. Absolution could be found in the church, a process that would strengthen its power and influence. Because the church could grant or withhold its absolution depending on a sinner's willingness to do penance in terms of the church's values and edicts, the church could enter into the most private spheres of life. Through the act of confession, the most intimate details of individuals' private lives were laid bare to representatives of the church—in effect, no behavior was free from intrusion from the outside.

Present-day conflicts over laws dealing with birth control and abortion in Western countries in part reflect the influence of religion and in part the lack of shared values regarding these behaviors. In the United States, much of the population has taken sides that coincide with opposing religious and ideological dogmas.[31] Yet religion alone cannot be implicated as the key contributor to such conflicts. Interestingly enough, similar conflicts occur in situations where the state officially and often vehemently rejects religious values, as in the USSR during the last seven decades and in pre-World War II Germany, where infertility became grounds for divorce in 1938.[32] The aim to increase the German population was the force underlying new legislation regulating marriage and divorce and, by implication, procreation.

Although relatively effective birth control devices have long been available to men, it was not until recently that devices have been available to women, which give them comparative safety in avoiding pregnancy when engaging in sexual practices. Not only have the women of most technologically oriented societies readily adopted these devices, but their use has become the subject of intense legal activity in the United States as well as in many other countries. As long as women could not rely on relatively safe and effective birth control techniques, their role within society was largely that of bearing and caring for children. It is likely that most women were happy in this role due to strong mother-infant bonds and the satisfaction they derived from nursing and raising their children. Over the last one hundred years this situation has changed dramatically. Modern medical advances have significantly lowered mortality among children. Women can protect themselves from pregnancy. As a result, women's roles in society are no longer limited to bearing and raising children. Nevertheless, their desire for equal economic and social rights still seems in part an illusion. Why? The answer may lie in the fact that women, unlike men, in fulfilling their sexual needs—an integral part of their desire for pair formation and pair-bonding—sometimes (intentionally or unintentionally) become pregnant.

The rights of women have come a long way and much of the social inequality of the past is now history. How much things have changed is illustrated in the 1857 Parliamentary debates in Great Britain on the reform of divorce laws prior to the adoption of the Matrimonial Causes Act.[33] It was stated that there was a profound difference between adultery committed by husbands and that committed by wives, because adultery committed by a husband does not cause "confusion of progeny." In these debates it was argued that this fact was an essential and important part of the offense, and thus adultery committed by a wife was "more criminal" than adultery committed by a husband. It is mind-boggling to contemplate the gap between the biological knowledge of today and the availability of biological information in the second half of the nineteenth century, shortly before the publication of Darwin's writings. Yet, relatively little has changed since then in the basic attitudes expressed in the laws of most societies, except perhaps in Western divorce and custody law.

As noted, effective birth control methods have enhanced the possibility of equal economic and political rights for women. They have increased women's ability to control the number and timing of births, and thereby exert control over the flow of property and the parental care of children. They have also made it possible to disassociate sex and reproduction. And, at least partially in consequence, new laws in the United States and other countries allow women to make important biological decisions, namely, avoiding continuation of a pregnancy when there are conflicting desires between the woman and the father of the child.[34] However, after the fetus reaches a certain point in its development, society, not the woman, determines if pregnancy should continue.[35]

Faced with the tremendously bitter conflict between Pro-Choice and Pro-Life groups, legislators, and judges, as well as the general public, might well take time to investigate statistical and historical data concerning the relationship between birth control and abortion. The incidence of abortion reflects the availability, quality, and use of contraception devices or medication. In the Soviet Union, where the quality of birth control is poor, we find the highest per capita abortion rate in the world.[36] Japan ranks third or fourth highest.[37] A disconcerting finding from the United States is that it ranks first among industrialized nations in teenage pregnancy and teenage abortion.[38] The teenage situation may be due to failures on various levels: educational, ethical (moral), and practical (exposing youngsters to situations and temptations at an age at which they are not mature enough to cope responsibly). This situation is aggravated by the lack of adequate contraceptive choices in the United States. The introduction of steroid oral contraceptives in the early 1960s made the pill the most popular method of birth control in the United States. By 1970 it was used by almost 10 million American women. Because of side effects and the physiological mechanisms by which the pill exerts its effects, it is not an ideal choice for all women desiring to avoid pregnancy.[39] Alternative forms of female contraception have been discussed by scientists since the mid 1970s. There have, however, been no new major developments with respect to contraceptives during the past two decades in the United States. The development had to occur in European research firms. The situation in the United States reflects the confusion that has developed because of court decisions and congressional actions,[40] as well as from litigation. It has become extremely costly for the pharmaceutical industry to develop or manufacture any type of contraceptive. Legal defense costs for the drug and insurance companies have escalated significantly. In addition, many scientific investigations have turned away from developing new contraceptives, in large part because the principal United States government funding agencies have provided minimal support for this area of research. (The study of infertility attracts more talent, probably because of better financial support.[41])

As intense and conflictual as arguments about sexuality and its control often turn out to be, it is wise not to lose sight of the fact that numerous alternative ways of limiting and encouraging population changes are known. Population-limiting devices include: delayed mating, celibacy, prohibition of adultery, monogamy, lactation and birthspacing by humans, and abortion. Population encouraging devices include: early mating, promiscuity, polygamy, concubinage, easy divorce, levirate, prohibition of population limiting devices, and special payments for offspring. Each legal system usually has a mixture of these institutions, one implication of which is that the debate about pro-life and pro-choice legislation hopelessly narrows the issues beyond the point where reasonable and intelligent debate is possible. We may hope that both Pro-Choice and Pro-Life groups will become more aware and vocal on these complex issues and espouse the aim of finding a common cause by providing women a choice of safe contraceptives in various forms (from IUDs to the day-after pill, which can prevent and interrupt implantation), in the context of the more complex issues that shroud sexuality.

There is little doubt that the growing decision-making power of women will influence present-day legislation concerning their familial and social roles more than it did at any previous time in the immediate past. Already it has had far-reaching effects on women's individual rights. They often find such rights in conflict with the concept of the family, couples as entities, the interests of the children, and at times even the interests of the state. This evolving situation will result in new conflicts concerning the rights of individual women, the best interests of the child, the interests of husbands (or sperm donors or sperm sellers), and the family.

Similar conflicts were the subject of the previously mentioned decision by the Supreme Court of Massachusetts in March 1973, which decided in favor of a 23-year-old wife. (This decision was confirmed in 1976 by the U.S. Supreme Court.) The wife's attorney argued that in this particular case the constitutional (fundamental) rights of the wife take preference over the rights of the husband: "the decision by the court to prevent the wife from having an abortion violates the constitutional (fundamental) rights of the wife, her right to make her own decisions, and deprives her of her freedom. It is, in effect, involuntary slavery." The husband's attorney argued that the husband-father has a right to a relationship with the child whom he had fathered: "We attempt to protect the relationship between father and child."[42] From a biological perspective, the decision is far from straightforward, however. Thus, it is not surprising that this decision as well as closely related ones continue to be the center of ongoing debate and conflict.

Biological facts cannot provide foolproof instructions about the rights of individuals in human family relationships. They may provide important insights, however. We may learn from observations of other species that each individual has social and biological roles and functions within the family unit. Moreover, if such functions are not carried out, the mental and physical health of children may be jeopardized. Children who grow up in families with both parents present are significantly less likely to suffer from mental disorders than those who grow up in families with single parents. In different cultural settings, these different family functions generally get translated into expectations about role functions in family units. Each individual is expected to act and react within a given range of behavior, to exercise specific responsibilities, and to do so consistently. Actions that *may* appear to an observer as exercises of power, or of overriding others' rights, can perhaps be traced to responsibilities implicit in intrahuman features of the individual's role. Individual rights in these cases are balanced by reciprocal arrangements or obligations.

As a general rule, man-made law regulating the child-parent relationship has at all times followed lines set down by biological fact. The biological mother has been recognized as such whenever she gave birth to an infant. In modern society, the child is legally assigned to its biological parents if both are known but, in almost all cases, to its biological mother. These practices contrast in important ways with the psychological and legal importance of biological fatherhood. This concept is relatively new in human social evolution. In various roles, adult males have been part of the family group since *Homo sapiens* emerged as a separate species. However, the relationship of males with the infant, child, or adolescent has depended on both the males' roles within the family group and male proclivities. Generally, bonds between father and child have developed if the father accepted the mother of the child as a member of the group or family in which he lived (or, in the case of matrilocal societies, if he went to live in her parents' household), if his family or group accepted her, or if he recognized and accepted a child as his progeny. The preceding divisions can be unsatisfactory, at least in those cases where they do not adequately reference the complexity and (often) the intensity of fatherhood. There is evidence supporting the view that fathers can assume most of the roles traditionally assumed by mothers (e.g., rearing and caring for offspring) and that they can perform these duties very adequately.

Today in most human societies the law assigns fatherhood to the husband of the mother. For many centuries in the West, the maxim *ubi matrimonium ibi pater* was a seemingly sound principle for legal decisions. In our times of sexual freedom this principle could have mixed advantages and consequences. To follow this principle now that biological fatherhood can be disputed or excluded to a high degree of certainty might have beneficial effects by stabilizing relationships between parents and children. Other trends can also be discerned, however, and it will take time before the full consequences of any changes in reproductive methods can be evaluated.

CARE FOR THE PROGENY

The role of biology in the rearing of offspring is older than the history of the human species, and the role of law in regulating the behavior of parents may be as old as law itself.

However, only very recently have scientists studied the biology of infant-rearing behavior. In discussing this subject, I will highlight some recent findings from studies of nonhuman primates where many of the new insights about parent-offspring relationships have been developed.

The mother-child relationship in nonhuman primates has been investigated across a variety of conditions, including: child-and-biological mother, child-and-surrogate mother, and child-and-adoptive mother. The experiments conducted by Harry Harlow[43] on rhesus monkeys, which were first published in 1959, indicate that bodily contact with the mother plays a primary role in the development of infant affection. A cloth-covered wire surrogate could partially replace the mother to young infants. As monkeys raised by cloth mothers matured, however, their behavior became abnormal: They sat and rocked in corners and were unable to interact with other monkeys in social encounters. When they reached maturity, they were unable to mate normally. If such females were impregnated, they gave birth. These "motherless mothers" tended to mistreat their infants, however, often to a degree that would have resulted in death had caretakers not intervened.

In 1969, 10 years after publishing "Love in Infant Monkeys," Harlow concluded from his long-term experiments that "there is no adequate substitute for monkey mothers early in the socialization (of rhesus monkeys)."[44] One may ask, why? Part of the answer has to do with infant development. Normal mother-infant relationships appear to catalyze the development of behavior traits in infants that are essential ingredients in subsequent mothering. Related research has shown that for normal development of infants, certain interaction events must occur at certain critical periods. Bonding between infant and mother needs to take place within hours after birth. Wire mothers, as well as replacing mothers by raising infants with peers only, are inadequate substitutes. Later, play with peers becomes critical, but normal play occurs only if mother-infant bonding has occurred first. Research by Bowlby and others[45] on human infants confirms Harlow's findings and underscores the significance of early infant-mother attachment and the long-lasting consequences of damage done when this attachment is compromised. Inadequate bonding leads to a predictable chain of events: Mothers who have been raised in atypical circumstances behave abnormally, infants engage their mothers in atypical ways, and mistreatment follows.

The preceding findings raise obvious questions about the changing roles of women in society and the possible effects of these changes on infant development.[46] Will greater rights of women be balanced by greater obligations to their infants? A sense of responsibility, one of the outcomes of human evolution, could influence the kinds of choices females will make given an increasing number of alternative options: career versus motherhood; postponement of motherhood until completion of education and establishment of a career; and part-time career and motherhood. This same sense of responsibility could also motivate employers, who often have to rely on the employment of a large percentage of female workers, to provide day-care facilities, and to institute nursing and visiting services to infants of working mothers. The problem of infertility and the various new technological ways of supplying infants to childless parents will also enter into these discussions.

Harlow's observations of motherless mothers among monkeys should encourage us to take a closer look at demands for social changes that may not take into account procreation and child-rearing methods. In this inquiry it is important to remember that children remained in close physical contact with their mothers until very recently in evolutionary history, especially during the early years of life. Infants seldom survived when they were separated from their mothers or a caretaking relative. Even to the age of four and older the separation from the mother often endangered the existence of children. Mother-infant bonding is not merely a custom or habit that can be changed at will without producing serious consequences. Human physiology evolved to function in the conditions that prevailed in early times. Drastic changes in child-rearing behaviors thus are likely to have undesirable effects on the development of infants and juveniles and lead to undesirable consequences in the next generation.

Today a growing number of women select careers which significantly reduces the time available to their offspring. Given this situation, one might argue that it is in the best interest of society to grant women the right to avoid motherhood or to limit the number of children without forgoing sexual gratification. It is true that controlling the number of offspring has been practiced for centuries through a variety of methods. Yet, the new contraceptive technology has greatly increased the safety and predictability of such controls and antiovulation medication, or medication that interrupts implantation and that is now available in Europe, is likely to be an even more predictable and safe method of contraception. When contraception fails, the individual and society have always been faced with decisions concerning the abortion of the fertilized ovum and, clearly, years of experience in dealing with these decisions have not significantly improved our ability to generate a solution satisfactory to all.

Our current situation leaves us with a number of unsolved issues. For example, what effect does abortion have on genetic fathers, already alive children, or the subsequent behavior of the woman who aborts? Are pair-bonds compromised? Does the unilateral decision to abort result in a general disregard for biological process and possibly legal behavior? There are also economic and the aforementioned child-care issues. Introducing a child into a situation in which there are inadequate resources for support and inadequate time on the part of the mother to develop mother-child bonds is questionable at best. In such circumstances, the overall personal and social benefits of abortion may far outweigh the negative ones. Yet, reproduction is central to human existence. It is, therefore, not surprising that many opinions are being voiced on this subject, and it is important that we continue to investigate the possible biological and social consequences of the change we are experiencing.

We now come to the father-child relationship. Because most nonhuman primate species do not appear to have a concept of fatherhood similar to that of humans (although some primatologists would dispute this point), we lack models to compare with and extrapolate to human fatherhood. For many nonhuman primate species, the human observer has great difficulty in determining the biological father—the father of a newborn in the group could be any one of a number of adult males. In such species, adult males (possible fathers) will readily defend an endangered infant and generally they are tolerant of infant and juvenile play. When there is only one adult male in the group, as with gibbons, marmosets, and titi monkeys, paternal behavior is seen. In the case of the siamang (a gibbon species), who engage in monogamous pair-bonds, the father carries the older infant during the day when the infant cannot as yet cover longer distances, and the mother carries the younger infant. At night the father will also share his sleeping quarters with an infant.

As previously noted, the importance of biological fatherhood in humans is a rather recent concept. To more fully understand the father-child relationship, a distinction should be made between the bonding process—which takes place between mother and fetus, and mother and infant—and the bonding a potential father may have for his child. The father's feelings may be strong, as may those of other parties with a genetic investment in the child such as grandparents and siblings. Certainly a father who is with the mother throughout her pregnancy, assists during birth, and later cares for and feeds the baby is likely to develop a strong bond with the infant. But even then it is not comparable to the mother-child bond developed during pregnancy and immediately after birth. The fact that the father has not actually carried the child and gone through the physiological changes associated with pregnancy means that the type of bonding that exists between the father and his infant will differ compared to that of the mother. Moreover, current findings suggest that the process by which bonds are formed between fathers and infants proceeds more slowly, and is primarily social or emotional. Thus, for the first days or weeks after birth, the tie between the infant and the male parent appears not to be as firmly established as it is with the female parent.

These differences in attachments have been examined in recent court proceedings involving surrogate mothers, most prominently in the recently tried *Baby M* case. In its 1988 decision, the New Jersey Supreme Court affirmed the rights of the birthmother (the "natural mother") and negated the previous adoption of Baby M by the biologic father and foster mother. In the area of adoption, the public policy of every state in the United States (with the possible exception of Wyoming) holds that consent to adoption prior to birth is invalid unless ratified after birth. This requirement acknowledges that the mother and unborn infant may have already established a preliminary bond. The courts of New Jersey have recently reemphasized that under the public policy of that state a mere private contract is not effective to renounce parental rights. Rather, consent may be withdrawn prior to formal adoption.[47]

No doubt, here again public opinion will be strong and influential when the various legal systems consider legislation to deal with the problems caused by these new developments. Public opinion, of course, is greatly influenced by the quality and availability of scientific information. Thus a critical question is: To what degree are legislators, judges, and the public aware of this information?

EVOLUTION OF THE HUMAN FAMILY STRUCTURE

Thus far in this chapter I have discussed and compared functions of primate sexuality and interactions between sexuality, bonding, and group structure. As noted earlier, laws tend to reflect our biological propensities and constraints; that is, the law distinguishes the various biological functions, and various laws tend to follow these functions. In the following section I will explore how these points apply to the human family structure, which is affected by our biological nature and the laws we devise to shape and constrain various features of sexuality.

Understanding the evolutionary history of the human family structure is a basic requirement for exploring many of the issues already discussed as well as problems that occur in the nuclear family, such as infertility, artificial forms of reproduction, divorce, and child custody. An evolutionary perspective is equally relevant for dealing with the extended family, where care of the infirm or aged poses a special set of problems; and to cope with conflicts in intrafamilial or extrafamilial relationships. The structure of the family and the various biological mechanisms that have led to family formation among humans are integral to the formation and cohesiveness of kinship groups or clans, and contributory to legal behavior concerning property rights,[48] moral obligations, and related issues that are addressed by law. One of the oldest written legal documents, the Law of Gortyn (400-500 B.C.), found on the island of Crete, attests to the fact that the issues addressed in this section are not new. This law, carved on stone tablets, treats property as a function of family support, favoring the women and children of the kinship group, and not as a function of individual rights.[49]

I will begin this section by developing some general concepts about the family. The data are taken primarily from anthropology and history and similarities between humans and nonhuman primates are noted. Perhaps nowhere in law are findings from biology more enlightening to legal thought than in family law. Likewise, nowhere in legal thinking are value judgments more deeply embedded than in views about the family. For this reason, biologically based information that might conflict with these values often is disregarded or rejected.

Because the family is fundamentally a biological group, and because it occupies such a central position in human society, any ethological exploration of legal behavior associated with family law must closely examine behavior within the family, beginning with the most obvious difference in human families: the monogamous family units of Western Europe and the polygynous family units present throughout parts of Asia and traditional Africa. Monogamy is by no means the most prevalent family unit. In two comparative studies carried out in the 1950s, anthropologists found that 84% of 185 studied societies or 75% of 554 studied societies practiced legalized polygyny.[50] Although contemporary anthropologists may question some of the criteria and statistical techniques used in the development of these findings, use of the most conservative and stringent statistical criteria does not negate the conclusions from these studies: The majority of studied societies practice legal marriage of one male to several females.

One explanation why families exist in different forms is found in the interrelationship between religion and family structure. Religious beliefs and practices have widely influenced marital and family law. Yet, our hominid ancestors lived in cohesive groups and formed family units long before modern religions began. For the most part, religions developed around already existing, conservative family structures and, like law, followed existing behavior. Religion, therefore, cannot be held primarily responsible for the differences found in the various family structures. More instructive insights may be found in historical reconstructions of early family units and in comparisons between human and nonhuman primate behavior. In all likelihood, mother-infant relationships were the cornerstone of these early family groups (certainly such relationships exist today among all known human groups). The mother with her infant was perceived as a unit, much as is the case among nonhuman primate groups. Moreover, across different environments, this basic bond between mother and infant is likely to have resulted in similar perceptions and values. The preceding points lead to questions dealing with the conditions and causes of different family structures.

Modern humans, like ancient humans—there is a difference of degree almost equal to a difference in kind—are shaping their environment through technology and, in turn, responding to environmental changes that are the work of previous generations. Living with these changes has required adjustments in behavior, including family structure. Behavior undoubtedly is the most malleable characteristic of all animals.[51] Its plasticity is not only a characteristic of *Homo sapiens,* but behavioral change has proven to be a necessity for survival. There is an obvious implication here: Given a species with adaptive behavior, and given that bonding structures change in relationship to environmental variables, then the assumption that one type of bonding relationship is optimal across widely diverse environments needs to be questioned.[52]

Environment alone does not provide a full explanation of behavior, however. Groups adapting to different environments develop different values with respect to family structure. The human species has had examples of monogamy, polygamy, polygyny, and occasionally even polyandry—a system of fraternal polyandry, with several brothers sharing a wife, is found in parts of India and in Tibet.[53] According to Eibl-Eibesfeldt, scarcity of food supplies and territory prompted the development of diverse forms of family structure.[54] He further suggests that polyandry was one way to achieve population control in an environment where each female needed several males to support her and her offspring. Other anthropologists disagree with this interpretation, primarily because polyandry does not provide a suitable basis for determining paternity, which, they argue, is critical in keeping males connected to the family and in control of most systems of family law. There are, however, established ways of dealing with paternity uncertainty. Patrilineality (*paterfamilias,* power over wives and offspring) often occurs in situations where paternity certainty is high and has the effect of maintaining social status and resource control, especially in stratified societies. Matrilineality has similar effects and is often present where paternity certainty is low. Child care, residency, power over offspring, and the transmission and control of property are involved here. Likewise, certain ecological variables apply. Large landholdings, for example, are usually associated with primogeniture. Where small groups are widely separated and often migrating, matrilineality is more likely to occur.

The family structure of our hominid ancestors may have shown familial varieties similar to baboon troops. Edey notes that:

> just as baboons have differing life styles, surely hominids did too, depending on where they lived and on just how severe the seasonal food- and water-getting problems were. Where such problems were at their worst, extra male hominids may be considered to have been just as expendable as surplus male baboons, and one-male family units may have resulted.[55]

There is of course no one-to-one translation between nonhuman primate data and the behavior of humans. Nevertheless, analogous functions and structures exist. For example, the various family structures among baboons discussed earlier imply a model of the family structure existing among our hominid ancestors who lived as gatherers. Among these human ancestors, the mother-infant unit certainly existed. Adult males probably were not actively occupied with parental tasks but functioned within the group as protectors of females and offspring. Among groups that engaged in hunting and gathering, food sharing became a necessity—certainly hunting could not become a principal feature of human social evolution until males (generally, the hunters) and females (generally, the gatherers) shared food and food-getting.[56] Once such sharing of tasks and their division between sexes occurred, the family (in its various forms) is likely to have become the most viable economic unit. (Generally, other primates do not engage in a high degree of food-sharing behavior. Some exceptions exist, however: sharing between mother and infant and meat sharing among chimps, which follows certain rules determined by the availability of meat.[57]) Of interest is the assumption that without the reinforcing effects of resource sharing, division of labor would not have developed. Studies of living hunter-gatherers indicate that a home base facilitates the division of labor. Women and children gather food nearby while men hunt. The family then reunites at the home base to share the results of their labors. It seems that for thousands of years, human social groups prospered when families had such a home base, which may be likened to the household in today's legal definition of family.[58]

Bonding between human males and females is not based on rigid mechanisms, as is seen for example among hamadryas baboons where males jealously guard and herd females. Like the hamadryas, however, human pair-bonds are not limited to females' time of estrous. Yet, the human female's continuous ability and capacity to participate in sexual activity and to provide physical comfort is likely to have been a critical factor in establishing and preserving stable pair-bonds, one by-product of which may have been to assure food and protection for the mother and her offspring. To survive, the human infant required constant, close maternal contact for a significant period of time, and it is likely that the human mother benefited from the help of males even in relatively favorable environments. The presence of the male as the father of the family instead of a mere sire became even more important when specialized skills had to be taught in order to assure survival of the progeny.

Although the human female's constant sexual receptivity attracted the male and presumably strengthened the pair bond, her continuous sexuality could also endanger the infant's life if early intercourse after childbirth resulted in pregnancy or disease. Pregnancy could terminate lactation. Disease could end life. In most primitive societies—even in those cases where it is reported that the causal connection between intercourse and pregnancy is not clearly understood—strong taboos regulate sexual intercourse before and after childbirth. The taboos are strictly observed among most polygynous families.[59] For this reason, polygynous structures may have proved adaptive when food sources lacked sufficient proteins for infants to survive without their mother's milk.

During the last 10,000 years, a relatively new life-style developed, based on agriculture. According to present findings, scientists postulate that modern humans learned to domesticate plants and animals in order to produce and store food supplies approximately 10,000 years ago. Excavations have generally placed the development of agriculture in the area of the Crescent Valley in the Middle East, where the oldest plant and animal fossils suggesting domestication have been found. (Recent evidence has suggested that agriculture may have developed independently in the New World.) From this (or several) area(s), the agricultural revolution spread to other parts of Asia, to Africa, and, eventually, to Europe.[60] During this spread, the typical family unit had probably developed to the point that extended families existed, comprising several generations and including one or more adult males and several females, infants, and adolescents, probably based on matrilineal descent.[61] In all primates there is a tendency to deal with non-family members in different ways and to show likes and dislikes.[62] One might assume that this tendency was strengthened by immobility associated with cultivation of land and the desire of individuals to possess the land (a predictable nutrient source) they had cultivated. They would defend it against those they disliked or did not know.

The agricultural economy relied on the labor of women, adolescents, and young adults. Although adult males may have spent time away hunting, a sufficient number needed to remain at home to defend the land against nomadic bands who might pillage food supplies or people. Defense needs limited the size of the territory, which in turn limited the amount of food available and the number of people who could be fed. Under these conditions the extended family, possibly monogamous but more likely polygynous, is likely to have been adaptive.[63]

Under these conditions two factors would have affected bonding: a relative increase in the time males spent near females and their offspring, and a greater familiarity with their offspring. As a result, stronger paternal bonds were possible. The father could teach his sons the tasks men performed. Possible precursors for such behavior can be seen among nonhuman primates. A relaxed and protective attitude by the adult males toward the young is a part of primate heritage. Adult male chimpanzees show great tolerance toward youngsters, even when they interfere with the males during copulation.[64] Wickler[65] describes the "infant schema" (a term coined by Konrad Lorenz), which points to certain characteristics of infant physiology and behavior that elicit broodtending responses. These responses are seen in both male and female humans.

Slightly more males than females are born in many human groups, and sex-related differences in death rates have been observed in certain areas.[66] In some groups, a number of adolescent or young adult males probably died while hunting for food or fighting against predators, while adolescent females led more sheltered lives. Hoebel points out that there would be approximately one and one half times as many females as males in a traditional Eskimo group if it were not for female infanticide.[67] In other groups, early or unsterile childbirth and lack of well-developed postpartum sexual taboos are likely contributors to a high death rate among young females. Depending on the ratio of young adult males to females, pair-bonding could have been institutionalized in the form of monogamy, polygyny, or even polyandry.

Beginning between 10,000 and 5,000 B.C., agriculture and animal husbandry increasingly became integral parts of man's economic environment. This significant change in the human life-style influenced family structure. The social order most likely to assure procreation and the raising of the young to adulthood in a horticultural situation is somewhat different from the structures that will be optimal in hunting and gathering societies. Initially, family relationships seem to have been channeled and regulated by taboos, commandments, and rules, many of which were based on religion. During these years human imagination had created myths and strong beliefs in the supernatural. The balance between incentives and constraints was molded by experience and the authority vested in leaders—persons of the highest status and dominance. In these settings, society became more structured and the family received the sanction of law and religion. This situation helped to assure reproductive success for persons of high status who would transfer wealth and status to their progeny. In some instances, adoption was used for this purpose, (e.g., in the Law of Gortyn).[68] Again, similarities to nonhuman primate behavior are striking. As among humans, status also can be transferred (or inherited) to a certain degree among chimpanzees. Infants of females who enjoy a higher status usually develop behavior that assures them higher status during adolescence and adulthood.[69]

An influential new economic factor developed about 5,000 B.C. The village-farming community, made up of 400 to 500 people, had been stabilized for at least 1,500 years. Then, a sharp increase in tempo saw these small communities grow in size and undergo a decisive change of structure. Communities became larger and they took on urban characteristics. In the Near East, they culminated in the Sumerian city-state with tens of thousands of inhabitants; elaborate religious, political, and military establishments; and extended trading contacts.[70]

This decisive cultural and social change was made possible by an agricultural economy efficient enough to produce surpluses. By 3,000 B.C., several important cities in lower Mesopotamia included a significant number of acres within their fortifications. The city of Uruk, for example, extended over 1,100 acres and may have contained as many as 50,000 people. In 5,000 to 8,000 years the human life-style had changed more radically than in the preceding 100,000 years. Humankind had learned to produce and store food instead of merely gathering and hunting it. This change released human energy for specialization and a whole spectrum of new activities. When cities reached the size of Uruk, crowding and increased contact with strangers may have affected physiological and psychological processes that influence aggressive and sexual behavior—certainly we know that crowding can do that among many species, including *Homo sapiens*.[71] Highly developed cults and ceremony among the newly urbanized population, as well as a strong belief in the supernatural, are likely to have provided outlets for repressed emotions and served to redirect aggression and strengthen latent inhibitions. The legal system of these social groups had to reflect the drastic changes in life-style. The rules and laws regulating their behavior probably took the form of religious commandments. Nonetheless, these rules dealt with legal concepts, for example, property, inheritance, rights, and obligations. It is possible that territorial demands changed the concept of property in regard to landownership. The urban life-style may have favored the concept of individual property rights that became part of the Western legal systems.[72]

Urban living conditions prompted a class system in which it was economically advantageous, and probably emotionally satisfying, to have a limited number of dependents. This may be one of the reasons monogamy became attractive and that some religious leaders[73] (who had emerged as a powerful class under the stratified social system of the Sumerian city-states) eventually proclaimed it as the only acceptable family structure. While this development undoubtedly took several centuries, and did not occur universally, it may explain the prevalence of monogamous family structures in Western Judeo-Christian society.

The monogamous family, whether arising by design or accident, proved adaptive. Strengthened by religious command, monogamy was the marital system that spread with Christianity to the entire Western world (much of which was already monogamous). Elements of the extended family remained. The care of the old, the sick or disabled, and widows and their dependent children remained a part of the family function well into the nineteenth century, and in some societies, such as rural communities, well into the twentieth century.

As technology, mobility, and modern urbanization again brought drastic and rapid changes, the family structure, especially in highly developed countries, shifted toward the nuclear family. This trend was accelerated by the increasing life span in the early twentieth century. Indeed, industrialization and exploitation of resources in the New World may have occurred as rapidly as it did because of the Western emphasis on monogamy, a family structure capable of greater mobility than the polygynous one. Christianity commanded that a woman leave her parents to follow her husband even to faraway lands. This fact perhaps partly explains why industrialization began and progressed more rapidly in Christian, and especially in Protestant, states where there was an emphasis on the individual's duty to prosper on earth in order to be rewarded in heaven.

Today, the ever increasing life span means that existing family structures often cannot accommodate the changes occurring over such long periods. Thus, it seems clear that the nuclear family has serious failings in the new social environment that is developing. It often does not adequately provide for the maturation and emotional support of individual family members. Rapid social change causes continuous conflict in sexual and family relationships. Demands for individual rights appear among individuals and social groups without the assumption of reciprocal obligations, partly because the services implied by such obligations are believed to be available from institutions such as welfare, children's protective services, and the health-care establishment. Whether the nuclear family will continue to prove adaptive may in part depend on the values expressed in present and future legislation, although the law is more likely to follow than to lead in the area of change. Although new structures of the family, such as the one-parent family and step-families, are receiving increasing legal attention, many individual needs are not being adequately met. Conflicts between individual rights and obligations are increasing. Society's dependents, the children, often are the victims.

Legal systems have developed various subdivisions to deal with the numerous conflicts relating to the family: The law of domestic relations, juvenile law, and adoption law are the traditional fields. Recently, legislation has addressed new problems resulting from modern biotechnological advances in the area of extrauterine conception or other forms of artificial reproduction. To have any hope of understanding the interactions between the evolutionary impact of biology, the technologically developed man-made changes in the environment, and the value judgments expressed in man-made law, it is necessary to address these bodies of information as a whole.

Assuming a very broad definition of the human family, one immediately finds that different disciplines use the same word, but their meanings differ. For instance, Black's law dictionary defines a family as "a collective body of persons who form one household under one head and one domestic government and who have reciprocal natural and moral duties to support and care for one another."[74] Although descriptively accurate, this legal definition of the human family does not incorporate the new insights into the meaning of family gained by the continuing research of anthropologists, sociologists, and other scientists. On the basis of anthropological and ethological data, *family* cannot be understood merely as a spatial concept expressed in the word *household*. A strong factor in the concept of family is its origin in the kinship group.

The most fundamental principle of the human family is that it is a kinship group. In most cases, it is spatially united in one household, but it certainly can take and has taken other forms, especially in Western society. A high percentage of children in the United States, for example, now live with a single parent or in a stepfamily. These children may have a variety of relationships in which roles and obligations are not clearly defined. Half-brothers and half-sisters from several marriages of both parents may live together. Step-siblings, step-uncles and aunts, as well as all the relatives from the original marriage, may continue to interact as a family. Family law must deal with the problems arising from this situation, yet the present legal definition of family does not describe the human family as a kinship group. The courts have had to address, for example, the visiting rights of grandparents to a grandchild living with a widowed stepmother who was divorced from the biological father. Or, consider all the complications arising from surrogate motherhood, open adoption, or other family-related options.

Family matters occupy a place of importance in all existing legal systems not only because of the biological importance of the family but also because families are integral to group life. As was noted earlier in this chapter, family structure varies in response to a variety of factors, e.g., historical, ecological, religious, and biological. These differences notwithstanding, family structures and intrafamilial obligations are defined and regulated wherever human families are found. Whether the definitions and regulations are derived from ancient taboos, religious commandments, ancient or modern laws, or contracts, definition and regulation are present. Moreover, family matters are the most frequent cause for judicial action in primitive societies and, in technologically advanced societies they are a major subject of legislation and adjudication.[75]

For decades, anthropologists and sociologists have been alarmed by what they view as increasing family disintegration, exemplified by divorce, violence, spouse battering, and child abuse. Many reasons for this disintegration have been advanced. Some theorists attribute this situation to changes in family types, or the increasing size and complexity of society, or rapid economic and technological changes, and to prolonged resource abundances. For others, disintegration is viewed as the consequence of rising expectations of what the family should provide for its members. From the perspective of behavioral ecology, abundance, predictable resource flows, and security reduce the need for cooperation in order to meet minimal daily requisites for food and shelter. Whatever view is taken, the usual results of disintegration—the neglect of children and the weak—lead to special legal considerations.

Legislation and commission reports of the 1960s and 1970s on divorce all emphasize the importance of family stability.[76] Opposition to easy divorce has often been motivated by the belief that such a policy will inevitably lead to the destruction of the institutions of marriage, family, and society. Is this belief correct? We can ask whether modern divorce laws actually foster family instability or whether they recognize the existing instability, or to what extent the law can effectively direct or channel behavior. To have a chance of answering these questions, it is necessary to focus on the way changes in laws regulating family relationships reflect changes in the behavior of people interacting with different environments. As environments change, behavior also changes, and history has shown that eventually the laws also change, reflecting changed human behavior.

Today we view breakdowns in basic family ties such as the pair-bond or the parent-child relationship to be one of the major signs of family instability. Although the causes of these events are complex and still poorly understood, there is little disagreement about the consequences. The constantly growing caseload in juvenile courts, the increase in violent, family-related crime and child abuse (increase of which may be real or exaggerated by a new group of professionals concerned with family affairs, but may also be at least in part a factor of inadequate reporting in earlier days), as well as mental disorders (which appear to be increasing), the neglect of the old, and related social ills, each have some ties to family stability and structure.

Given the preceding points, it is not surprising that there are enormous pressures for reform of domestic relations laws. For example, since the 1960s, state legislatures and federal courts have been particularly active in revising these laws. In the early 1960s, the United States Supreme Court undertook to reinterpret existing laws to accommodate conflicting new demands. In *Griswold v. Connecticut* in 1965, the Supreme Court based its decision granting married couples the right to use contraceptives not on the Fourteenth Amendment, but on a "right to privacy older than the Bill of Rights."[77] In *Eisenstadt v. Baird* (1972),[78] the Court granted single persons the right to use contraceptives. In *Roe v. Wade* (1973),[79] the Court granted a broad right to obtain abortions and, in 1976, the Court decided that a married woman could not be required to obtain the consent of her husband for an abortion during the first trimester of pregnancy, and said that "the State may not impose a blanket provision . . . requiring the consent of a parent or person *in loco parentis* as a condition for abortion of an unmarried minor during the first 12 weeks of her pregnancy."[80] In general, the decisions of the United States Supreme Court have continued to recognize the conflicts between the various movements advocating differing positions on rights and responsibilities of couples (married or unmarried), especially concerning abortion. At the present time this kind of adjudication is being vehemently attacked. The focus of the attacks is the 1973 *Roe v. Wade* decision. The outcome of this confrontation seems uncertain.

CONCLUSION

Ethology provides us with a ray of hope for clarifying many of the areas of conflict discussed in this chapter. Ethologists recognize fully the problems involved in the fact that modern men and women shape their environment through technology to a degree unprecedented in human history. Their findings demonstrate that humans are capable of adjusting their behavior, including family structure, to adapt to the changes that new technology creates. Many ethologists hold that behavior is the most malleable characteristic of all animals, and plasticity of behavior applies as much to bonds and family organization as it does to an individual acting alone. The conflict between individual and societal interests, however, is at the bottom of most of our problems, and the absence of a general agreement on what constitutes the good of the group is a major obstacle in the way of society's ability to adapt and survive.[81]

NOTES

1. In the United States as in all Western nations the law has considered any form of polygamy (or bigamy), whether entered into for religious or other reasons, not only illegal but criminal. In the United States in 1788 the State of Virginia enacted the statute of James I, death penalty included. In 1862 Congress made bigamy a crime in the Territory of Utah, where polygyny was practiced by the population adhering to the Mormon religion. There are still numerous cases of this practice in Utah and even recently such offenses have again been tried in court. (See Fox, R. N.d. *In the matter of plural marriage, Reynolds v. USA 1879* and *Cleveland v. USA 1946.*)

Another example of how the game is played by different rules in different religious settings is the use of ostracism. While the exclusion of certain members from the social group is practiced universally as a cultural solution to hysteria and a means of protecting the cohesiveness of the group, the justification given for these actions and the rules applied in these situations vary depending mainly on the prevailing religion or ideology (see Gruter and Masters, *Ostracism,* supra note 30, Chapter 1).

2. With the changes in sexual mores and the increase in the number of unmarried couples living together, various conflicts arising in these situations have been brought to court. In *Marvin v. Marvin* the Supreme Court of California, in 1976, held that the plaintiff was entitled to equitable relief, not by the provisions of the Family Law Act but on the basis of an implied contract, agreement, or joint venture. (The court stated that the provisions of the Family

Law Act do not govern the distribution of property acquired during a nonmarital relationship; however, the court could enforce express contracts between nonmarital partners except to the extent the contract is explicitly founded on the consideration of meretricious sexual services, despite contention that such contracts violate public policy. The plaintiff was successful in obtaining support payments—palimony.) There are numerous publications on this subject and the various decisions in similar cases dealing with property rights, parental rights and other conflicts among unmarried couples. For example, see Galante, M. A. 1986. "Courts not wed to 'palimony': Who can recover?" *The National Law Journal* 45 col in v8 (July 14):3 col. 3; Reidinger, P. Family law: Cohabitation confers property rights. *ABA Journal* 72 (Oct l):86 92(2).

There have also been a number of law suits concerning custody by unwed parents. Here the courts are following the "best interest of the child" doctrine. See Reuben, R. 1988. Justices to weigh custodial rights of unwed parents. *Los Angeles Daily Journal* 30 col in v 101 (Nov. 28): pl col 4.

3. Wickler, supra note 35, Chapter 1.

4. Kummer, supra note 2, Chapter 3.

5. Wilson, supra note 29, Chapter 1.

6. Lorenz, K. 1966. *On aggression*. New York: Bantam Books.

7. Beach, F. A. 1965. *Sex and behavior.* New York: E. Wiley. Sexual behavior in nonhuman species has been studied by observation of actions, but it has also been thoroughly investigated in its interrelationship with the nervous system, hormones, and individual experience. See also Washburn and Hamburg, supra note 29, Chapter 5, at 279. According to Washburn and Hamburg, when animals are closely related, it is more likely that the internal biological mechanisms on which their actions are based will be similar.

8. van Lawick-Goodall, J. 1971. *In the shadow of man*. Boston: Houghton Mifflin.

9. According to Wickler, the function of sexual behavior can change, and pair-bonding mechanisms can be used to redirect aggression. Wickler, supra note 35, Chapter 1.

Konrad Lorenz described this pattern in cichlids and greylag geese. Lorenz, K. 1963. *Das Sogenannte Böse*. Wien, Austria: Borotha-Schoeler.

10. Chimpanzees also use display behavior when confronted by natural events that seem to arouse feelings of frustration. For example, at the beginning of the rainy season they perform very impressive "rain dances" (an expression used by Goodall). They obviously dislike rain, but in spite of their intelligence they have never erected any shelters that protect them from rain (according to recent reports, orangutans will do this).

In general, chimpanzees try to avoid contact with water; they use stepping stones when crossing streams. Often their hair stands out while they do this, as is the case during displays. Chimpanzees will also display in front of waterfalls. Goodall suggests that the use of displays in this context could be seen as a rudimentary form of worship similar to the worship of natural forces practiced by early human groups. From these precursors the first religions may have evolved.

11. Address by Jane Goodall, Stanford University Lecture Series, in Stanford, California, October, 1973.

12. Hall, E. T. 1977. *Beyond culture*. Garden City, NY: Doubleday.

13. Washburn, S. C. and D. Hamburg. 1972. Aggressive behavior in old world monkeys and apes. In *Primate Patterns*, ed. P. Dolhinow. New York: Holt, Rinehart and Winston.

14. Masters, W., and V. Johnson. 1970. *Human sexual inadequacy*. Boston: Little, Brown. In their clinical studies of the physiology of human sexuality, Masters and Johnson found that sexual response varies in intensity and frequency among individuals and among age groups, but only by degree; the absence of sexual response is amenable to treatment. The symptoms of human sexual inadequacy are impotency and orgasmic dysfunction; they can be caused or aggravated by fear, pressure, and ignorance. The researchers, estimating that one-half of all American marriages are threatened by sexual dysfunction, attribute the cause of marital breakdown to a dysfunction of the pair-bonding element in sexual behavior.

Other medical publications also emphasize the importance of sexuality as a pair-bonding element and view sexual dysfunction as the leading cause of marital discord. They call for greater educational efforts by medical schools to close the glaring gap in the sexual knowledge of the medical profession.

15. An estimated 50 million abortions—half of them illegal—are performed around the world each year and as many as 200,000 women die from them. (Study by Jacobson, J. L. *The global politics of abortion*. Worldwatch Institute, July 25, 1990.)

16. Most of these regulations are old and contained in the civil code of the various states. There have been some changes due to the increasing number of women who have earnings from outside work and the earlier maturation of children who become independent at an earlier age.

17. During the last few years higher numbers of child abuse or physical abuse of women by their husbands have been reported and tried. So far this increase has been attributed to social and other environmental changes in modern society. The number of reported sex offenses, particularly against children, has risen dramatically. In 1976 an estimated 7,500 cases of child sexual abuse were reported, while in 1983 nearly 72,000 such cases were brought to the attention of the authorities. (Finkelhor, D. 1986. A sourcebook on child sexual abuse. Cited in the *Journal of Psychiatry & Law* Summer 1988.

Straus, M. A., and R. J. Gelles. 1986. Societal change and change in family violence from 1975 to 1985 as revealed by two national surveys. *Journal of Marriage and the Family* 48 (August):465-479. Statistics indicated decreased violence in 1985 compared to 1975 both in violence directed against children as well as against spouses, mainly wives. These conclusions were attacked in an article by T. Stocks (1988. Feedback. *Journal of Marriage and the Family* 60 (February):281-291. There is, however, a consensus that family violence, especially reported sexual abuse, has increased dramatically during the last years compared with data of the 1960s. Prosecution of these acts has proved to be extremely difficult. Lately the trend has been toward the repeal of sexual psychopath statutes that concentrated on the right of the defendants in the disposition of cases, and the expansion of legislation that attempts to improve the probability of successful prosecution of child sexual abuse cases and ease the difficulties of child victims in the courtroom.

18. See generally Capron, A. M. 1987. *Alternative birth technologies: Legal challenges. University of California—Davis Law Review* 20:679.

19. 17 *Encyclopaedia Britannica* 1008 (1964) sv. mode of reproduction.

20. Sorenson, R. 1973. *Adolescent sexuality in contemporary America*. New York: World. See Table 45, p. 385.

21. Chimpanzees, for example, have long-lasting mother-infant ties, and the chimp family, consisting of a mother and her offspring, often continues together even when the offspring reach maturity. Yet, copulation between a physically mature male and his mother has not been observed, although infants are sometimes seen to mount and thrust on their mothers when the mothers showed sexual swellings. Copulations between known nonhuman primate siblings have been recorded, but in the one instance, the female tried to escape the attempts of her two brothers to mount her. Similar indications of a mother-son incest taboo have been observed among the Japanese macaques. Research involving rhesus monkeys also indicates an inhibition regarding mother-son mating. Rhesus males rarely mate with their mothers, in part it appears because sons retain inferior, infant-like attitudes toward their mothers. This reflects a specific inhibition that is independent of rank, thus restricting males even when they are socially dominant over their mothers. In hamadryas baboons, father-daughter mating apparently is reduced by the daughters being carried off by young adult males before they are of sexual interest to their fathers. van Lawick-Goodall, supra note 8. Kummer, supra note 2, Chapter 3.

22. Lévi-Strauss, C. 1969. *The elementary structures of kinship*. Boston: Beacon.

23. Fox, R. 1971. *Kinship and marriage*. Baltimore, MD: Penguin Books.

24. In the State of California, the civil code deals with incestuous marriages in paragraph 4400 of the chapter on Void Marriage, declaring marriages between parents and children, ancestors and descendants of every degree, and between brothers and sisters of the half as well as the whole blood, and between uncles and nieces or aunts and nephews as void from the beginning whether the relationship is legitimate or illegitimate. This law derives from a regulation enacted in 1872.

The Penal Code of the State of California paragraph 285 states: "Persons being within the degrees of consanguinity within which marriages are declared by law to be incestuous and void, who intermarry with each other, or who commit fornication or adultery with each other, are punishable by imprisonment in the state prison." This law derives from a regulation enacted in 1850.

25. Winick, C. ed. 1970. *Dictionary of anthropology*. Totowa, NJ: Littlefield, Adams, 280, 281; Seymour-Smith, C. ed. 1986. *Dictionary of anthropology*, Boston, MA: G. K. Hall, 147.

26. Feierman, J. R. ed. 1990. *Pedophilia*. New York: Springer.

27. Boswell, J. 1979; repr. 1953. *Life of Johnson*. Ed. R. W. Chapman. Oxford, UK: Oxford University Press.

28. Gazzaniga, supra note 2, Chapter 1.

29. Malinowski, supra note 7, Chapter 1.

30. Essock-Vitale, S. and M. T. McGuire. 1980. Predictions derived from the theories of kin-selection and reciprocation assessed by anthropological data. *Ethology and Sociobiology* 1:233-243.

31. See Glendon and Calabresi, infra note 35.

32. Ehegesetz vol. 6.7.1938 (RGB 1.I: S. 807).

33. Boswell, supra note 27; Gruter, M. 1944. *Die Stellung der Ehefrau im englischen Scheidungsrecht*. Doctor of Jurisprudence diss., University of Heidelberg, Heidelberg, Germany; Howard, G. E. II. 1904. *A history of matrimonial institutions*. Chicago: University of Chicago Press.

34. See *Bellotti v. Baird*, 428 U.S. 132 (1976), motion to vacate order denied, 97 S. Ct. 251 (1976); *Planned parenthood of Missouri v. Danforth*, 428 U.S. 52 (1976).

This opinion is stated explicitly and vehemently in numerous influential books of the modern feminist movement. Friedan, B. 1963. *The feminine mystique*. de Beauvoir, S. 1953. *The second sex*.

35. In her 1987 book (*Abortion and divorce in western law*. Cambridge, MA: Harvard University Press), M. A. Glendon discusses the fragmentation of the community and its values concerning women's rights to abortion and compares the American situation with pertinent legislation in other Western countries. Interestingly enough she predicts —what has started this summer—the demise of the *Roe v. Wade* ruling, which she calls narrow. "Perhaps—when and if Roe self-destructs—out of its wreckage can be devised a set of guiding principles for state legislators. . . ." This seems to be happening now. She also sees the first woman appointee to the U.S. Supreme Court, Justice O'Connor, as playing a critical role in this decision. C. Gilligan addresses this problem in her *In a different voice*. 1982. Cambridge, MA: Harvard University Press. The availability of choice, and with it the onus of responsibility, has now invaded the most private sector of the woman's domain. . . . For centuries, women's sexuality anchored them in pacifity, in a receptive rather than an active stance, where the events of conception and childbirth could be controlled only by a withholding in which their own sexual needs were either denied or sacrificed.

G. Calabresi in *Ideals, beliefs, attitudes, and the law*, supra note 17, Chapter 4, also discusses the emotional dispute embodied in the *Roe v. Wade* case in his chapter "When ideals clash." He states "that courts have . . . consistently held that the male involved does not have rights in the matter indicates again that the key issue is not one of fetal life, but of women's rights to equality."

There are many more learned books written expressing various opinions on this subject. It is probably the most emotionally driven conflict of our time. Law will have to deal with it, and currently the U.S. Supreme Court is doing this by turning some of the decisions over to the individual states, thus avoiding a centralized showdown of the opposing forces.

36. Tietze, C., and S. K. Henshaw. 1986. *Induced abortion: A world review*. New York: Guttmacher.

37. Djerassi, C. 1987. The politics of contraception—The view from Tokyo. *Technol. Soc* 9:157.

38. Trussell, J. 1988. Teenage pregnancies in the United States. *Fam. Plan. Perspect* 20:262.

39. Djerassi, C. 1989. The bitter pill. *Science* 245:28.

40. U.S. Congress, Senate Subcommittee on Monopoly of the Select Committee on Small Business, Hearings on Competitive Problems in the Drug Industry (91st Cong., Washington, DC, 1970), vols. 1-3 (Oral Contraceptives).

41. Djerassi, C. supra note 39.

42. Belotti v. Baird, supra note 34.

43. Harlow, H. 1959. Love in infant monkeys. *Scientific American* 6:68.

44. Jolly, A. 1972. *The evolution of primate behavior*. New York: Macmillan.

45. See Bowlby, J. 1969. *Attachment and loss, Vol. 1: Attachment*. New York: Basic Books; Hassenstein, B. 1973. *Verhaltensbiologie des Kindes*. Munich, Germany: R. Piper, concerning the significance of the early infant-mother attachment and the long-lasting consequences of damage done during this developmental process.

Also see MacDonald, K. B. 1988. *Social and personality development*. New York: Plenum; MacDonald, K. B. ed. 1988. *Sociobiological perspectives on human development*. New York: Springer.

46. Lancaster, J. B., J. Altmann, A. S. Rossi, and L. R. Sherrod. eds. 1987. *Parenting across the life span: Biosocial dimensions*. New York: Aldine de Gruyter.

47. Supreme Court of New Jersey, *Baby M 537 A2d 1227 (1988) 109 NJ 396*; see also *In the matter of Baby M*, Amicus Curiae Brief of the Gruter Institute. 1988.

48. Cooter, R. 1990. *Inventing Property: Economic Theories of the Origins of Market Property Applied to Papua New Guinea*. Olin Foundation Working Paper #88-5, University of Virginia. Cooter applies economic theories to explain the origins of market property in the kin groups of Papua New Guinea. He describes a present-day situation in which property rights of land are regulated by customary law that evolved out of kinship behavior; he suggests that theft and systematic mismanagement of Indian land in the United States and Canada may by the result of imposing policy which is not compatible with evolved behavior (customary law) in the kinship groups of tribal people.

49. Bücheler and Zitelmann, supra note 8, Chapter 2.

50. Beach, supra note 7; Johnson, R. 1972. *Aggression in man and animals*. Philadelphia: W.B. Saunders.

51. Wickler, supra note 35, Chapter 1.

52. Wahl, E. 1973. Influences climatiques sur l'evolution du droit en Orient et Occident. *Revue International de Droit Comparé*. The author traces the development of different concepts of property rights in the East and West to climatic and economic differences that have shaped "the image of man." Similar conclusions can be drawn regarding adaptations of human relationships.

53. Fox, supra note 23.

54. Eibl-Eibesfeldt, supra note 2, Chapter 2.

55. Edey, supra note 44, Chapter 1.

56. DeVore, supra note 1, Chapter 3.

57. Teleki, G. 1973. The omnivorous chimpanzee. *Scientific American* (January):33-42.

58. *Black's law dictionary*'s definition of a family: "A collective body of persons who form one household under one head and one domestic government and who have reciprocal natural and moral duties to support and care for one another." 728 (4th ed., 1968).

59. Malinowski, B. 1929. *The sexual life of savages*. New York: Harcourt, Brace & World.

60. Braidwood, R. J. 1960. The agricultural revolution. *Scientific American* 9:131.

61. There might have been some forms of matriarchy among these early groups. Greek mythology certainly has inspired such perceptions, and mythology often is based on some kind of historical data. Bachofen has dealt with this subject extensively. See supra note 15, Chapter 1.

62. This phenomenon has been observed also among nonhuman primates:

> Although a primate has access to all members of his group, he usually shows marked preferences for some members while hardly ever interacting with others. Such subgroup formation is partly based on unexplained individual affinities and partly on kinship and age-group preferences . . . Nonhuman primates recognize only matrilineal kinship . . . The subgroup of a mother and her children shapes (their) social relationships. (Kummer, supra note 2, Chapter 3.)

63. Malinowski, supra note 7, Chapter 1, and Hoebel, supra note 33, Chapter 1. Both argue that polygamous households are advantageous because they assure the availability of food for infants for sufficiently long periods of time, as well as facilitate adherence to postpartum sex taboos.

64. van Lawick-Goodall, J. 1968. The behavior of free living chimpanzees in the Gombe stream reserve. *Animal Behavior Monographs* 1:161.

65. Wickler, supra note 35, Chapter 1. Infant Schema. pp. 255-265.

66. Ziegenfuss, W. 1956. *Handbuch der Soziologie*. Stuttgart, Germany: Ferdinand Enke Verlag.

67. Hoebel, supra note 33, Chapter 1.

68. The Law of Gortyn, supra note 8, Chapter 2.

69. van Lawick-Goodall, supra note 64.

70. Adams, R. M. 1960. The origin of cities. *Scientific American* 9:153, 160, 161.

71. Washburn and Hamburg, supra note 13.

72. R. Cooter, supra note 48, describing the society of the Papua New Guinea and their concept of property.

73. These early religious leaders in the cities of lower Mesopotamia are believed to be the first group to be freed from subsistence labor. They manifested their power in the temples, which were the largest and most powerful institutions of these early cities. Adams, supra note 70.

74. *Black's law dictionary* 728 (4th ed., 1968).

75. During the last 10 years the law had to grapple with various new definitions of *family* in housing and zoning situations. Many communities have laws restricting housing according to varying definitions of what constitutes a family; see for example, the "living-in-sin" zoning code in Denver, Colorado, similar regulations in Poughkeepsie, and Chicago where it applies to public housing projects. Supporters of these regulations argue that it encourages strong family-oriented neighborhoods, helpful in the fight against drugs. In Massachusetts a case has been pending before the Massachusetts Appeals Court for six years, involving three unmarried couples who were barred from public housing. Other rulings concern single parents or gay couples. In West Hollywood, California, a 1985 domestic partnership ordinance gives unmarried couples, heterosexual, gay, or lesbian, the same rights as married couples. All these conflicts concern not the "family" as such but the "common household" part of the present legal definition. By adhering to this interpretation the concept of family is even further removed from the kinship tradition. Yet in King William County, Virginia, two sisters, Kim Taylor and Cam Porter, members of the

Pamunkey Indian tribe, are attempting to change a tribal law barring them from the reservation because they both married white men.

76. Rheinstein, M. 1972. *Marriage stability, divorce and the law.* Chicago: University of Chicago Press; The California Governor's Commission on the Family, Report 9 - 11 (1966).

77. 381 U.S. 479, 486 (1965). Also see Bender, S. 1974. *Privacies of life,* (March) Harper's, 63.

78. 405 U.S. 438 (1972).

79. 410 U.S. 113 (1973).

80. *Planned parenthood of Missouri v. Danforth,* 428 U.S. 52, 74 (1976). See also *Bellotti v. Baird,* 428 U.S. 132, motion to vacate order denied, 97 S. Ct. 251 (1976).

See also *Akron v. Akron Ctr. for Reproductive Health,* 462 US 416, 76 Ed 2d 687, 103 S Ct 2481 (1983) and *Richard Thornburgh v. American Coll. of Obst. & Gyn.,* 476 US 747, 90 L Ed 2d 779, 106 S Ct 2169 (1986).

81. Hardin, G. J. 1968. The tragedy of the commons. *Science* 162:1243.

SIX

Ethology and Environmental Law

The preceding discussion on ethology and family law built on historical data and prehistoric findings, and on what we today know about the evolution of the human family. It focused primarily on explanations provided by the rapidly increasing knowledge of biology to enhance our understanding of the family and, in turn, to guide our interpretations and revisions of family law. Knowledge developed over the past four decades in biology and related fields (anthropology) has significantly altered our views of the family. To the degree that this knowledge is introduced into the public domain and becomes a part of legal thinking, significant changes in family law can be anticipated—formal law will be more effective if it follows more closely the living law. Readers are likely to disagree with some of my interpretations of findings, perhaps also with the degree to which I have argued that evidence from studies of nonhuman primates leads to insights into human behavior. These differences notwithstanding, it is difficult to refute the basic theme of the last chapter, indeed of the entire book: namely, evolution has resulted in a species (*Homo sapiens*) that is strongly motivated to accomplish specific biological goals, that acts in predictable ways to achieve these goals, and that is constrained from engaging in many behaviors.

This chapter cannot rely to the same degree on previous experimental or documented events in our past, yet the influence of our evolutionary history again comes into play. The species-characteristic behaviors discussed in the preceding chapters turn out to have important implications also when applied to the type of environmental law to be discussed, namely, laws designed to preserve the environment over the long term. Many of the behaviors which humans are strongly predisposed to pursue result either directly or indirectly in the waste of important resources and in environmental deterioration; and there are evolved constraints on behaviors which would prevent further exploitation. For this reason, ethological studies are equally important for conclusions and predictions leading to new policies and legislation in this field. They are the most reliable guidelines to use in solving ever-mounting environmental problems and conflicts.

The first part of this chapter addresses the following question: What species-characteristic behaviors are most relevant to environmental law? The second part of the chapter addresses the question: What are the implications of these behaviors for laws dealing with environmental preservation?

SPECIES-CHARACTERISTIC BEHAVIORS AND THE ENVIRONMENT

Species-characteristic behaviors are behaviors that have been selected in the remote past because they favored an adaptive advantage. They are called species-characteristic because they are influenced by genetic information and because, given normal maturation, they are enacted by the great majority of members of a species. As noted in Part I, among adults the frequency of these behaviors is in part determined by events during development, in part by cross-person differences in genetic makeup, and in part by environmental constraints and options. Although these factors explain some of the variance, and although it is critical to incorporate them into our thinking about environmental preservation, it is equally important not to lose sight of the fact that strongly predisposed behaviors occur and that they occur with great predictability and regularity. Within limits, predisposed behaviors can be shaped (frequency increased or decreased), but they will not disappear. Consider, for example, the behaviors of acquiring, controlling, and utilizing physical resources (e.g., property, water, precious metals, etc.) for personal and kin benefit and competing for social positions that are associated with greater access to resources (e.g., high status). A significant part of recorded history documents the good and bad social and individual consequences resulting from such behaviors. We have wars and we murder; we marry, deceive, cheat, and change ideologies and religions to alter or preserve resource access and control. There have been legal and educational attempts to moderate these behaviors, to redirect them into more culturally acceptable behavior, or to constrain certain benefits that accrue (e.g., tax those who acquire excessive resources). Nevertheless, these behaviors persist. Moreover, it appears that a large percentage of people are more satisfied with the results of such behavior than they are with results of alternative behaviors. Recent events in Eastern Europe can be interpreted from this perspective. They suggest that the failure of socialist economies is due in part to the unwillingness of humans to devote more than a certain proportion of their personal effort for general social benefits at the expense of individual material success and status. The turn from a socialist to a market economy signals a turn from social to individual resource-acquisition goals.[1]

What are the species-characteristic behaviors that are of greatest interest for environmental preservation laws? Most of the crucial behaviors have been discussed earlier. These behaviors will be briefly reviewed from the perspective of environmental law.

SELF-INTEREST

A fundamental axiom of biology is that organisms act in their own self-interest. This axiom applies across the animal kingdom, from protozoa to humans. Self-interested behavior can be both direct and indirect. People work hard, follow rules, manipulate and deceive, and bide their time to achieve self-interested goals. Moreover, self-interested behavior of two or more individuals often leads to conflicts. Conflicts may escalate to warnings and threats. Among humans, there are a variety of known ways to resolve such conflicts (compromise, compensation, apology, and behavior change). Yet, these options often go unutilized and conflicts persist, at times to the point that participants' lives are endangered.

Self-interested behavior is strongly motivated and exceedingly difficult to eliminate.[2] For environmental law, the relevance of this point is the same as it is for all of law: Laws that do not take into account the strength and importance of self-interested behavior will fail to achieve their desired ends (e.g., shaping environment-preservation behavior).

NEPOTISM

People act nepotistically. They invest in their offspring and kin. They make special rules that apply only to family members. They leave property to offspring and relatives. At times, they deceive and exploit others to enhance economic and social advantages for family members. The predisposition to act nepotistically is a second axiom of biology that has direct relevance to environmental law.[3] Efforts to curtail such behavior by social means, group rules, or formal laws have been only marginally effective and any effort at long-term curtailment would necessitate intense and continual monitoring.

The importance of this axiom for environmental law is that nepotistic behaviors are likely to continue no matter what laws are developed. There are varying degrees in this behavior, however. It is conceivable that laws can be developed that constrain environmentally destructive nepotism (e.g., family controlled lobster-fishing territories in New England)[4] provided two conditions are met: such laws are largely restricted to environmental issues and do not attempt to constrain all forms of nepotism; and laws facilitate the acquisition of direct and indirect benefits to individuals and the kin of those who obey laws.

Consider, for example, the problem of the *tragedy of the commons*.[5] As the scenario goes, one or a few persons graze their cows on the commons and the cows thrive. Seeing this, others start to do the same thing. Eventually, there are too many cows for the available food and space. The cows become undernourished, milk production falls, and the commons is ruined. In this example, the importance of the commons is twofold: it references *public property*, and thus is analogous to much of our environment (e.g., air, water, parks, streets, etc.); and the products (milk and meat) of grazing cows impacts the lives of grazers and their kin. Several possible resolutions exist. Individuals can negotiate between themselves concerning who will keep how many cows on the commons for what period of time. This resolution can work if there is a limited number of individuals and cows and if there is a negotiation formula on which participants agree, which is most likely if the grazers are kin or at least members of a face-to-face community. A second solution is to restrict the use of the commons to a few individuals and their cows—for example, on a first-come, first-serve basis, as is often done to limit access to choice fishing spots. This solution may work if there are alternative direct and indirect benefits to those (and their kin) who are prohibited from grazing their cows, if there are either unused common resources (other fields for grazing, as in the opening of the West), or alternative benefits. In the latter instance, benefits might include inexpensive milk and dairy products as well as increased time available for alternative economic endeavors among nongrazers. A further requirement is some type of market system that mediates economic interdependence between grazers and nongrazers. In addition, some means of enforcement needs to be in place. A third way is to assign rights (use of the commons for grazing) and obligations (limited number of cows per area) to separate parts of the commons and let people buy and sell their rights and obligations.[6] Often called privatization, this works best if the commons can be divided into private property. However rights to the commons are defined, some mechanism of enforcement is essential, and the solution presupposes an ideal market economy, which, for a whole variety of biological reasons (e.g., xenophobia, trust, territoriality) never quite exists.

Each of the solutions above may be analyzed in greater detail. Each analysis leads to the same general conclusion: Solutions are possible only within a more complex structure that accounts for a variety of needs and behaviors. Much of this structure is provided by the legal system.[7]

RECIPROCITY

People engage in non-kin reciprocal behavior and build social-support networks.[8] They help others because the reciprocal benefits they are likely to receive in return usually are unavailable among relatives. Reciprocal behavior has certain risks, however. The person who receives help may refuse repayment. Although the tendency to engage in reciprocal relationships is a strongly predisposed behavior, the predisposition may not be as strong as it is for nepotistic behavior. Reciprocal arrangements work when individuals can recognize those who give and receive benefits[9] and there are social consequences for not reciprocating, such as socially ostracizing the person who repeatedly cheats or defects.[10]

INDIRECT RECIPROCITY

The importance of this axiom for environmental law is that many of the solutions to problems of environmental law require that persons engage in reciprocal behavior. As noted above, humans appear to be less strongly predisposed to this type of behavior than to nepotistic behavior, although among persons known to each other those who are asked to help usually do so and those who are helped usually reciprocate. It is a very different matter, however, when one moves from reciprocal obligations that prevail in familiar social situations to the type required in environmental law: In the vast majority of instances reciprocators will not know each other. I will refer to this type of reciprocity as *indirect reciprocity*[11] For example, if I recycle materials from my home so that they can be used again, I have no immediate impact on the camper who decides whether to pollute a small mountain stream (and vice versa). Yet, for environmental preservation to work, indirect reciprocity is essential—the camper and I must cooperate.

Related issues have to do with reciprocity-related cultural traditions and laws. In the United States, if a hiker falls and is injured, and another hiker finds him, there is no legal responsibility requiring the noninjured hiker to provide aid. That aid is usually provided is not only one of the "laws of the trail," but also a manifestation of our predisposition to assist those who are in need of help. In other legal systems of Western countries, aid-giving of this kind is prescribed by the law and, under certain conditions, the failure to provide assistance can be punishable. In these countries, the formal law has followed the living law, whereas in the United States the absence of such laws may reflect our deeply held beliefs concerning individual rights and freedoms. The preceding example contains different elements compared to situations where reciprocators are unknown to each other. Face-to-face contact takes place between the hikers and some degree of initial bonding is likely to occur even if help is not provided. To the degree that indirect reciprocation is a necessary condition for successful environmental preservation, the likelihood of such reciprocation occurring is far less than when two hikers meet. For indirect reciprocation to occur, two psychological steps are essential: intensification of our sense of ourselves as participants in the welfare of others, for example, through acting to preserve the environment;

and a sharpening of our capacities to engage in acts of environmental preservation.

Two other points, not quite axioms, can be included in the general category of species-characteristic behavior. The first concerns the tendency of primates to exploit and destroy their environment. Most nonhuman primates are highly destructive of their environment.[12] Such behavior may not be intentional. Rather, most primate species appear to have evolved under conditions in which environmental destruction did not have long-lasting consequences. Thus in their search for food, primates destroy mangroves, remove bark from trees, uproot trees, pull out roots, destroy birds' eggs, and so on. When their environment no longer yields sufficient nutrients, they move to new environments. In the meantime, the environment they have just left begins to undergo self-replenishment. More extreme and lasting forms of destruction are seen among humans. Sometimes destruction is intentional (e.g., wartime bombing of Britain and Germany, asphalting nutrient-rich farmland, destruction of the South American rain forest), although there are usually a host of reasons justifying such behaviors. Sometimes destruction is unintentional (e.g., disposal of nuclear wastes without understanding the consequences, *Exxon Valdez* hitting a rock). There are attempts to reduce the frequency of both intentional and unintentional environmentally destructive occurrences. In some instances, these attempts have been successful. The Great Lakes, for example, are considerably less polluted than they were a decade ago, and there are fish swimming in many streams in New England for the first time in three decades. Nevertheless, the environment continues to deteriorate.

It is in the area of environmental destruction that economics often enters as a key variable. It is easy enough to identify a single polluting lumber mill on a small river in a remote area of the world and to write and enforce a law that prohibits pollution. It is another matter, however, when one attempts to lessen environmental destruction and increase environmental preservation via cleaning the air in a city like Los Angeles. There are multiple contributors to polluted air. Solutions often require major changes in industrial procedures and the purchase of expensive new equipment, as well as changes in the daily habits of those who pollute. Solutions also often result in economic hardships, such as unemployment, losses on investments, price increases in commodities, and so on. In turn, cities and states are required to invest resources in unemployment payments and for the relocation of unemployed persons. As a consequence, solutions are often delayed. Exemptions are granted. Time frames are extended. And conflicts increase between those who have made efforts to preserve the environment and those who have not or who have minimally complied. Most important, the destruction of the environment continues and preservation efforts are delayed.

The case of air pollution in Los Angeles is instructive because environmental destruction has continued, indirect reciprocity has failed, and smog-reduction laws have lacked clear incentives. In 1970, local air quality control officials estimated that Los Angeles would be smog-free by 1976. By 1980, the estimate had been extended to 1985. Minor gains in selected areas of smog control were made each year. Yet, in 1990, the smog is worse than it was in 1970, and the decision has been made to start again. In the interim, cars, incinerators, factories, barbecues, old trucks—literally anything that smoked or released fumes—have been implicated. The current perception is that cars are the primary source of air pollutants. In response to this perception, the public has been asked to drive less, particularly at critical hours. The response to this request is that there are more vehicles on the roads of Los Angeles than ever before. Indirect reciprocity has failed. Increasingly stricter laws are likely as are increasing legal efforts opposing the laws. Moreover, because of the lack of incentive features in smog-control laws, attempts at circumvention are likely to be common.

The second point deals with short-term goals. There is little evidence to suggest that as a species we act on other than short-term goals.[13] Parents certainly invest in their children for benefits that will not be gained for several decades. We even invest in our grandchildren. But this is vastly different from the willingness to invest now for what might occur in the year 2500 and after. Investment in the distant future based on indirect reciprocity, which benefits unknown strangers as well as kin, has no basis in our evolutionary history. The counterargument, that things that were begun centuries ago and from which we still reap benefits, such as the building of universities and the development of legal systems, does not refute the general correctness of this point. These institutions had to have short-term benefits for some, even if only those with power. And even if it is granted that such behavior was undertaken with full realization of its long-range implications, such behavior is as often the exception as it is the rule.

The implications of the preceding points seem clear enough: any environment-related laws that are likely to achieve their hoped-for impact and that are likely to be followed for a reasonable period of time need to be conceived within the behaviors, motivations, and constraints our species brings to our interaction with the environment.

BIOLOGY AND LAWS DEALING WITH PRESERVATION OF THE ENVIRONMENT

I will begin this section by addressing the following question: Does the history of agriculture imply that we have the capacity to engage in environmental preservation? It is reasonable to argue that agriculture could serve as a model of land preservation and that the model could be generalized to other environment-preservation problems. Agricultural land is utilized to produce products, it is replenished either naturally or artificially, and it is used again for production. Leaving aside the issue of chemical fertilizers and the impact of these chemicals on the soil, organisms within the soil, and nearby environments, the question above can be given a biological perspective in the following way: It is unlikely that any significant natural selection has taken place since humans began agricultural behavior, circa 10,000 years ago. That is, it is unlikely that the experience of our recent ancestors with agriculture has eradicated the genetic basis of human social behavior described in the preceding chapters. This view has to be balanced against features of our history which clearly indicate that as a species we have the ability to change our way of life and to readjust goals when advantages are recognized.[14]

Clearly, the introduction of agriculture resulted in major changes in migration patterns, the utilization of alternative living strategies, and various forms of territoriality. The more important points to address, however, concern the survival, feedback, and economic factors associated with agriculture, for these factors provide insights into both the potential and the constraints of the agriculture model.

Much of the motivation underlying agricultural efforts has been associated with personal and kin survival. In situations in which our own or our kin's survival depends on the food we produce, we are likely to consider preservation important for the food-producing area of the environment. At the same time, agriculture was historically associated with the emergence of large-scale communities, irrigation systems providing collective goods, and the centralized state. This was evident in the emergence of the Ancient Near Eastern civilizations along the Tigris and Euphrates, and similar events that occurred in pre-Columbian America.[15] Thus, to the degree that environmental preservation is motivated by short-term personal and kin survival considerations, the agriculture model may be a valuable guide to environmental preservation. Potential applications of this model are difficult to think of for urban areas, however. To return to Los Angeles and its smog, parents have been aware for years that vehicle emissions are in part the cause of increased respiratory infections among young children, possibly also of increases in other lung diseases, such as emphysema and cancer. Nevertheless, everyone drives. For the present, the connections between personal and kin survival and environmental preservation in urban areas do not appear to be compelling enough for most persons to change behaviors that clearly are environmentally destructive.

An equally relevant point concerns the relationship between agriculture and feedback. In nearly all agricultural activities the products of one's tilling, fertilizing, and planting efforts are seen within a few months. There are exceptions of course. Fruit trees take several years to mature. Yet even they, as well as such trees as cork oak, which take 50 years to yield satisfactory harvests (and then take 10 more years before their bark can be stripped again), grow from year to year. Although agriculturalists of recent centuries have been willing to tolerate a greater length of time between planting and reaping the products of their efforts, the period is still well within the memory of those who participate in these activities. To the degree that environmental preservation requires efforts in the present to reap rewards generations hence, the agriculture model does not appear to provide much help, although again, there may be elements on which we could build.

There is the economic issue. Because of the costs of land, water, fertilizers, and labor, agriculture is increasingly tied up with economic constraints. Excess produce needs to be sold to offset the costs of production; land needs to be mortgaged to purchase seed and fertilizer and to pay workers. The system of agriculture is supported and shaped by a massive economic and political infrastructure which at times is helpful to agriculture, at times destructive. Yet the key point of agriculture remains the same: One expends certain resources in advance on the expectation of receiving certain rewards in the future. Again, the turn-around time of these events is relatively short. Thus, to the degree that environmental preservation requires expenditures in the present for benefits to be achieved after the lifetime of the expenders or their offspring, the agriculture model offers minimal help. Besides, with agriculture, too, success in the enterprise has been at the cost of enormous consequences in off-site pollution.[16]

From the perspective being developed here, the issues of laws dealing with environmental preservation turn out to be something like the following: To preserve and improve the environment it is necessary to engage in specific behaviors now; these behaviors are often costly, both in terms of time and of resources; many of these behaviors will have minimal discernible effects (benefits) within the lifetime of the person enacting the behaviors. Is *Homo sapiens* capable of such behavior? I will attempt to answer this question by discussing some of the suggestions for environmental preservation that others have made in the context of biology and law. An interesting outcome of this inquiry is that the combination of indirect reciprocity and laws designed to reinforce changes that provide incentives for the economic and educational systems appears to offer the most promising way of achieving the kinds of behavior changes that are essential for environmental preservation to succeed.

PUBLIC EDUCATION

Public education means informing the public of the consequences of behaving or not behaving in certain ways. Certainly this approach is possible and, increasingly, it is one of the methods used for shaping behavior.[17] As a rule, educational efforts are most effective with the young who are less constrained by economic necessities. In turn, they often influence the behavior of their parents. For behaviors such as excessive water use, gathering certain types of materials that are increasingly rare (e.g., redwood), using household materials that damage the environment (e.g., paint thinner for oil-based paint), or littering, this method has been effective.

There are limits to public education, however. For example, in its current form, education appears to be an ineffective way of reaching the goals of reducing the following: the number of automobiles, the use of the majority of oil-based products, the felling of rare trees, the number of people living in highly dense environments, single occupant cars, the slaughter of whales (by certain countries), and so on. The limits of effectiveness appear largely to be consequences of economic and other types of self-interest. Lumber companies wish to stay in business. General Motors strives to build fast cars, and a part of the economic well-being of the United States and many of its citizens is tied to the number of cars that are built and sold.

And, some of us like to sleep later than others, which makes car-pooling inconvenient.

LAWS WITH HARSH PENALTIES, TEMPTING AWARDS, AND ATTRACTIVE SUBSIDIES

Each of the alternatives (harsh penalties, tempting awards, and attractive subsidies) has been tried and, in one form or another, each represents an economic approach. Each has been successful in specific areas. Laws with harsh penalties now prohibit disposing of certain types of materials except in designated areas. When laws are violated, the violator is responsible for the cleaning up process and may incur fines. Awards are given to individuals, groups, and institutions that most closely achieve certain environmental preservation goals. Subsidies are sometimes provided to companies or others that refrain from engaging in certain acts.

The limitation of this approach is that after initial enthusiasm it has proven difficult to sustain these behaviors. Responses to stiff laws, competition for awards, and subsidies are often effective for short periods but generally they are not self-sustaining. Even laws that seem excessively harsh, such as those that increasingly apply to the oil-transportation industry, do not quickly result in acceptable solutions (e.g., double hulls on oil tankers) on the part of the industry and violations are constantly identified. Again, self-interested economics may be implicated.[18] A related issue, that of the continuing excessive use of oil products, appears to be a clear example of environmental exploitation by the producers coupled with the illusion among users that an alternate energy source will appear before such products no longer exist.

COMMUNITY-SUPPORT SYSTEMS

Community-support systems associated with social ostracism can be a means of influencing those who don't comply.

This approach has proven to be effective in communities where the degree of social cohesion is sufficient for the public to understand that there are undesirable social consequences for those who do not comply with informal rules or laws. Thus individual communities will set up self-imposed recycling and nonlittering practices and often achieve high degrees of compliance among participants. The short-term goal of participants is community acceptance for engaging in behavioral change. Of course, restraining behaviors by the use of social ostracism is very effective in face-to-face societies,[19] but the larger the community, the less reliable this method becomes.

This approach seems to work as long as the community maintains a certain character. In California, for example, community support systems have worked well when the community membership has been stable.[20] Systems have worked poorly, however, where community membership fluctuates. Both direct and indirect reciprocity can be implicated. For direct reciprocity, time is required for reciprocal relationships to develop, for a new arrival in a neighborhood to make friends and to adjust to neighborhood values. These events need to occur before a person is likely to conform to community expectations.

AUCTIONING OFF THE ENVIRONMENT

One of the familiar ideas designed to limit environmental exploitation is that of "auctioning off environmental pollution." This idea assumes that people are willing to compete with each other over who has the right to pollute (or will pay a sufficiently high fee that others will tolerate their pollution). In principle, if a limit is set on the total allowable pollution, pollution could be contained. Moreover, resources paid for the right to pollute could be used for environmental replenishment.

At least three types of limitations exist.[21] High degrees of monitoring are required to assure compliance; economic inequities are inevitable; and, most importantly, at the present time the idea is politically unacceptable. Company A may buy the right to pollute and, when payment for the right is included in the cost of its product, Company A may produce a product for Z dollars. Company B, in complying with environmental protection regulations (because it doesn't have the right to pollute), will have to expend resources (e.g., purchase new equipment) to compete successfully and, for this example, the cost of the product for Company B will be set at Z+ dollars. A likely consequence, therefore, is that Company B will reduce its compliance to achieve short-term economic goals. A more serious problem concerns the side-effects of privatization. Citizens may feel that if the rich can buy the right to pollute, the average individual has no obligation to protect the environment.

REDUCED COMPETITIVENESS, LIMITED ENVIRONMENTAL DENSITY, AND IDEOLOGICAL/RELIGIOUS SOLUTIONS

A variety of alternative approaches have been suggested, and in some instances they have worked. Reducing economic competitiveness and therefore, presumably, the need to pollute the environment to retain a competitive advantage, is one such suggestion. This suggestion seems possible only if there is a high degree of government regulation. In principle, this type of economic solution is already in use with certain types of farm subsidies. Eventually, such systems backfire for self-interested reasons: namely, individual interests turn from limiting production to exploiting subsidy options (e.g., purchasing land on which one did not intend to produce so that subsidies can be received).

Limiting population density is another alternative. Some have said that the explosive growth in the world population of humans assures the defeat of every environmental protection strategy short of limiting the reproduction of *Homo sapiens*.[22] Some environments are difficult to preserve simply because they are overutilized—a human case of too many people on the commons. Reducing density (e.g., encouraging people to move elsewhere) has the effect of lowering usage. It only delays the problem of environmental exploitation, however, and generally does little for preservation. In particular, this solution can only work if population itself is controlled. Otherwise, those who practice restraint are exploited by migrating incomers from areas without comparable resources. The logic of the tragedy of the commons applies to the children of the farmer as well as to his cows. Zero population growth on a global scale may be the most important single factor in environmental protection.[23]

A third alternative deals with ideological and religious movements that are associated with environmental preservation. They offer interesting examples. From one perspective, they are highly effective in changing the behavior of those who participate in such movements. The Sierra Club and the World Wildlife Fund are examples. Such groups receive their primary opposition from those who suffer or stand to suffer economically from accepting the policies of such groups. Effective as these groups have been, they address only a limited number of issues. For example, environmental preservation is not confined to purchasing a piece of land and limiting its use, except from the short-term view. Events occurring outside the particular piece of land may have a significant impact on the land in question and render the effort of preservation relatively ineffectual (e.g., Chernoble).

CONCLUSION

Our evolutionary path has resulted in a far from ideal species, particularly with regard to our ability to address and master the environmental challenges that currently confront us. Numerous preservation suggestions have been put forth and tried. Many offer hope, but in a limited way. Any one, by itself, is not adequate to preserve the environment, in large part because of conflicting behaviors, many of which have their origin in our species genome.

We come then to the following situation: We are considering ways to shape the behavior of a species that has a phylogenetic history of exploiting its environment for self-interested reasons, that acts primarily on short-term goals, engages in nepotistic behavior, and is cautious about direct reciprocity, even more cautious about indirect reciprocity, and rarely engages in indiscriminate altruism. What might be done and what place should law have in the process of shaping behavior?

The most important changes that are needed involve education, encouraging greater indirect reciprocity, and a change in economic policies and priorities.

In-depth education is needed in schools and universities, and by the news media, that deals with the short- and long-term consequences of certain behaviors, the short-term benefits of preserving the environment, and alternative ways of achieving important goals. To start with, one might think of 50 years of such education so that at least two generations of *Homo sapiens* are familiar with the issues. An important focus of this effort should be as follows: behavior now is an investment in as-yet-unborn kin and an improved quality of life for kin; this is the benefit received for efforts in the present. Eventually, such education will have to address population size.

A second area concerns efforts to change our behavior toward greater direct and indirect reciprocity. The key here is to develop a view that as individuals we have obligations to unknown others and they to us. It is obvious enough how direct reciprocity laws might apply to the hiker who finds an injured person and is required by law to help. It is perhaps less obvious how preserving water in Northern California will be of value to the Eskimos. Here, indirect reciprocity is involved. To have a chance of working, indirect reciprocity requires that we alter different environmentally destructive behaviors in different environments. It requires that the relevance of the changed behaviors to life in other environments be understood. Thus in Northern California, the most important short-term goal might be to reduce the amount of water used because of the adverse consequences of overuse on species diversity. For the Eskimo, the most important short-term goal might be to discontinue killing seals because of interference with the biology of the sea. A further implication of this approach is that short-term goals will have to be changed frequently because the environment changes. The law will play an important role in the failure or success of such efforts. Because of conflicts between self-interested behavior and behavior that indirectly advantages others, laws will be essential both to shape and guide behavior and to punish those who do not comply with the law. One implication here is that a much more flexible and knowledgeable legal system (e.g., knowledge of the implications of reducing species diversity) is required compared to systems that exist presently. Another implication is that there will have to be incentives for indirect reciprocity. Moreover, the recommendation that new types of laws that would reward indirect reciprocity be developed is likely to be effective only to the degree that there is extensive education about the laws, their functions, and the compliance of others.

This brings us to change in economic policies and the dilemma faced by trying to balance different types of market economies with restraints on short-term self-interest and competition. We now know that the type of market economy that has prevailed in the United States has been and continues to be associated with significant and in many instances irreparable destruction of the environment. We are also aware that the tendency to exploit the environment for self-interested goals, be they individual or group (e.g., corporation) goals, has not disappeared. Witness, for example, the continuing legal disputes over access rights to certain natural resources, or opposition to obvious environmental protection efforts. We are also aware that environmental exploitation and destruction has had even worse results in countries guided by socialist concepts. Economic solutions alone thus are unlikely to cure the relationship between human nature and the environment. What seems possible at this time is a shift in economic policy to allow significant incentives for environmental preservation. To cite but one well-worked example, corporate taxes might be discontinued for 5 years for the first major automobile producer to develop a smog-free engine. The point here is to acknowledge that self-interest, which translates into economic incentive, needs to be channeled into environmental preservation efforts and away from short-term economic reasoning that does not take into account the environmental consequences of behaviors used to achieve success.

Such an approach would rely heavily on law, an informed government, and an informed public. Laws would have to be timely and addressed to the most critical environmental issues. The outcome would have to be monitored and, depending on their success or failure, laws would have to be changed without long delays. An informed government means that those who make and administer laws will have to be significantly more informed about species-characteristic behaviors and the environment than is now generally the case and that they would be motivated to use this knowledge in their decision making. Among other things, this means that they would have to understand the basic insights of biological (ethological) studies. Only if environmental protection becomes part of the living law, the law expressed in human behavior, will formal law achieve its goal.

NOTES

1. Flint, J., et al. eds. 1990. *Tearing down the curtain*. London: Hodder & Stoughton.
2. Aronson, E. ed. 1981. *The social animal*. San Francisco: W.H. Freeman.
3. Hamilton, W. D. 1964. *J. Theor. Biol.* 7:1-64.
4. Reader, J. 1988. *Man on Earth*. Austin: University of Texas Press, Chapter 6.
5. Hardin, supra note 81, Chapter 5.
6. Cooter, R., and T. Ulen. 1988. *Law and economics*. Glenview, IL: Scott, Foresman.
7. Dorfman, R., and N. S. Dorfman. eds. 1977. *Economics of the environment*. New York: Norton.
8. Trivers, supra note 7, Chapter 2.
9. Axelrod, supra note 10, Chapter 4. Rodgers, W. H., Jr. 1986. The evolution of cooperation in natural resources law: The drifter/habitue distinction. *University of Florida Law Review* 38 (Spring):2.
10. Supra note 30, Chapter 1. *Ostracism*, eds. Gruter and Masters.
11. Alexander, R. D. 1987. *The biology of moral systems*. New York: Aldine de Gruyter.

12. Krebs, J. R., and N. B. Davies. 1987. *An introduction to behavioral ecology*. Sunderland, MA: Sinauer Associates.

13. Dawkins, P. 1986. *The blind watchmaker*. New York: Norton.

14. Dubos, R. 1980. *Man adapting*. New Haven, CT: Yale University Press.

15. Wittfogel, K. 1957. *Oriental despotism*. New Haven, CT: Yale University Press; Hirth, K. 1984. Xochicalco: Urban growth and state formation in Central Mexico. *Science* 255:579-586. On the connection between evolutionary biology, agriculture, and the origin of written legal systems, cf. Masters, R. D. 1989. *The nature of politics*. New Haven, CT: Yale University Press, Chapters 5-6.

16. E.g., National Academy of Sciences, Board of Agriculture. 1986. *Pesticide resistance: Strategies and tactics for management*. Washington, DC: National Academy Press.

17. Roper Poll. 1990. *New York Times* (July 16): "A recent opinion poll, for example, showed that college educated Americans are more likely to support environmental preservation than those who have not gone to college or not finished high school."

18. Polinski, A. M. 1989. *An introduction to law and economics*. Boston: Little, Brown.

19. Gruter, M. 1986. Ostracism on trial. In *Ostracism*, eds. Gruter and Masters.

20. Ellickson, R. C. 1986. Of Coase and cattle: Dispute resolution among neighbors in Shasta County. *Stanford Law Review* 38:623; Coase, R. H. 1960. The problem of social cost. *J. of Law & Economics* 3:1

21. Hahn, R. W. and G. L. Hester. 1989. Where did all the markets go? An analysis of EPA's emissions trading program, 6 *Yale J. Reg*. 109.

22. Ehrlich, P. 1990. *Population explosion*. New York: Simon & Schuster.

23. Ibid.

Epilogue

This book has taken us from general concepts and ideas, wisdom acquired during recorded history, through parts of biology, to the application of these ideas and findings in family law and environmental law. While the written word covers scarcely 200 pages, it has been a long, rather exhaustive journey. At this point I will pause and look back at some of the highlights of the journey.

I realize that the preceding chapters have done little more than scratch the surface of an exceedingly complex and demanding subject—the relationships between biology and law in all their facets are still unexplained in the sense of detailed prescriptions to be used by legal actors. From what is known at this time, from the insights gained in 20 years of study, I feel more qualified in offering an opinion on what is not going to work compared to pinpointing methods that will be likely to work. Suggestions of what will improve the situation, choosing the less harmful or better alternative—this is within our grasp. I will close this book by addressing some lingering issues. And, looking to the future, I will identify some areas of law where ethological studies are likely to inform our understanding of law and its functions.

One of the main themes of the preceding chapters was that many features of man-made laws can be traced to the rules (predisposed, learned, or a combination) that influenced the behavior of early humans and their ancestors. Although we do not have, and probably cannot have, detailed facts about the social life of the earliest homonids, it is possible to make certain assumptions based on archaeological discoveries and through comparisons of present-day human behavior with the behavior of nonhuman primates.

In taking this approach, a picture emerged in which the behavior of the individual and the group gradually developed together and refined capacities to follow rules necessary for survival. During the course of thousands of years, these rules developed into commands, taboos, and rudimentary laws. As societies increased in size and complexity, many commands, taboos, and rudimentary laws came to be written laws. This approach led to the conclusion that legal research not only can, but ultimately must, attend to the theories and hypotheses of biology in formulating inquiries about the relevance of proposed laws, the predictability of legal behavior, and the effectiveness of existing laws. The reasoning leading to these conclusions is straightforward, and in my view it seems difficult to refute: Family-environment interactions determine the shape and organization of social groups; and adaptation of groups takes place through changes in the behavior of the individual group members or, more precisely, in the individual's functions within the group. These behavior patterns, although largely "culturalized" among humans, are activated and at the same time constrained as a consequence of our biological nature.

Results from a multitude of studies point to three factors that are primary in the development of group structure and the emergence of group rules: kinship (family attachments), reciprocal relationships, and hierarchy. Culture hides the fact that these are critical—perhaps the most critical—factors underpinning human society and social order. Only when kinship relationships are clear and reciprocity systems among non-kin are established has it been possible to achieve more remote, but still biologically based, organizational principles, as occurs, for example, in complex contracts and the concepts of property.

Kinship references a series of biological attachments that both spring from and lead to mutual dependence. In kinship systems, the claims of one party are regarded as the obligations of the other. Person-to-person, person-to-group, and group-to-person attachments are central to these systems and in any association deprived of these attachments, due to a lack of reciprocity or other causes, deterioration of the systems will occur. When attachments are necessary for survival, as in the case of kinship attachments, they are strongly supported by drives based on neurochemical processes as is seen, for example, in sexual attachment and mother-infant bonding. When arrangements such as indirect reciprocity are necessary for (non-kin) survival, we are faced with a more complex and fragile situation, however.

At some time during the evolution of man-made law, in an attempt to solve the conflict between individual self-interest and the welfare of the group, a social machinery had to develop in order to resolve conflicts inherent in group life.[1] When enforcement or restoration in kind was not possible, breach of contract and disrespect of property rights (as well as other actions detrimental to the social order) had to be dealt with, usually by punishment. Punishment of individuals who did not comply with the rules could bring the desired result—group order—only because punishment, like compliance, can yield a resolution to conflict and lead to a sense of security within a group. The effect of punishment can be "rewarding" if it constitutes an atonement and serves to reinstate an erring individual to his or her position within the group. Among chimpanzees, for example, the "law enforcer's" threatening gesture is followed by the deviant's submissive gesture, which generally wins from the law enforcer a reassurance gesture and which signifies a reinstatement of normal relations—one might almost say "atonement."

As central as the family is to multiple areas of law, and as many changes as we have seen in the allocation of family-related rights during the last one hundred years, lingering issues remain. To step back in time for a moment, it is worth noting that through the centuries women were not only granted rights of custody to the infants they bore, but were also held responsible for the well-being of the infants whenever the norms of the society imposed obligations to keep every or certain newborn human infants alive.[2] Mothers, not fathers, were brought to trial when abandoned infants were found. Famous dramas, like Goethe's *Faust* (1808), as well as reports of spectacular trials, indicate the enormous importance given to the obligations of the mother.[3] To this day, an abandoned newborn infant arouses the interest of the community. The incident is reported in the news media, and in most instances a search for the mother is begun. If the mother is found, she is brought before a court.[4] Even if she is no longer subject to the extremely cruel sentences of past centuries for her behavior, she is likely to suffer penalties. Rarely has public opinion held the father responsible. Especially where the abandonment concerned illegitimate children, a search for the father and an attempt to prove him guilty have rarely entered into the consideration of the courts.

Given these views toward the obligations of mothers and fathers, it would seem to follow that, together with the responsibility for the welfare of the newborn allocated to the woman, society should also grant her the rights to custody and care of the child if she so desires. But is this view applicable any longer? Increasingly, fathers raise their children and increasingly when families split up there is mutual responsibility for the care of offspring. Modern technology (artificial milk) and child care add further complications in that they free mothers from infant caretaking tasks they previously could not avoid. An obvious question is: Where is biology in this changing picture? The answer lies less in the technical details of keeping offspring alive and more in the organization of the family. It is this unit that strongly influences a child's growing sense of right and wrong. Strong family bonds, for example, affect the growing child's acquisition of respect for possession. The components or determinants of social interaction—kinship, hierarchy, and conflict resolution—have their origins in family relationships. Thus in considering the changing character of the family, it is critical that we keep in mind features of biology other than mother-infant bonding. While the mother may be the critical person in the initial bonding of an infant, the father has an important role in the socialization of the child, in refining systems of reciprocity, and in developing law-abiding behavior.

The points that were addressed in family law extended themselves into the area of environmental preservation. The consequences of changing family laws may bring rewards or new problems a generation or so later. This can be frightening and compelling indeed when we think of the well-being of our offspring. Yet when we come to environmental law we are faced with time frames spanning hundreds of years. We are required to invest now for the benefit of generations living in a future our mind can barely conjure. It is difficult to empathize with persons living in such a distant future, at least to the extent that we make sacrifices for them.

Looking at our own life and expecting rewards for our conduct, most of us, to a lesser or greater degree, turn to philosophy, religion, or just plain faith. Religion taught us to wait for the life-ever-after—heaven and eternity—to find rewards for altruistic behavior during our lifetime, our good deeds. For some of us, exposed to the new findings in genetics, it might be possible to follow this direction. We may include in these rewards which we do not live to see during our life on earth, and which will come to us after death, benefits that will be enjoyed by our offspring or the progeny of others with whom we share the genes from a common gene pool: Something left of us on earth will benefit in the distant future.

Apart from this, how can the law deal with the present dilemma of a deteriorating environment and our seemingly unsuccessful attempts to halt its continuing deterioration? Leaving aside situations in which we have been ignorant about the consequences of our actions, environmental protection can be narrowed down to the fundamental conflict between individual needs and the interests of the group. There is nothing new about these conflicts. They have always existed. Moreover, they are present even in the most primitive human societies[5] where only rudimentary forms of legal rules exist. The more complex a society, the more difficult it seems to find even temporary solutions to these conflicts, yet solutions are necessary for the survival of the species. To find solutions will require significant changes in our behavior, particularly in the area of indirect reciprocity. The family may be far more critical than we have realized; interactions with remotely related family members may provide the initial model for indirect reciprocity. Without such experiences, there is little to build upon.

As I have emphasized throughout this book, one of the functions of law is that of balancing the rights and obligations of the individual and the group. Such a balance of mutual expectations is the basis of group life. Nowhere is this point more relevant than in the area of environmental preservation. The law must provide a yardstick by which individuals become capable of predicting the consequences of their actions. Insights gained into the interaction between law and behavior relevant to the family are likely to be useful in judging more complex issues in larger social groups. Any effort to change the social order through law are *ipso facto* attempts to change behavior and it is in the family setting that the first major attempts to change behavior occur.

Turning to areas of law that have not been discussed in this book, property and contract need to be mentioned as parts of the legal system that offer a fertile field for ethological analysis. As we have seen, a rudimentary form of behavior that resembles the human respect for possessions has been observed in chimpanzees and baboons. Hamadryas males have a respect for possession of females once a pair-bond has been established. The chimpanzees use a begging gesture that would seem to imply a certain respect for possession, especially if a dominant animal begs from a less dominant one, and even more so if it is refused. This type of behavior—a sort of proto-respect for the possessions and claims of others—probably existed in early humans. The degree of respect for the possessor depended on the situation and the readiness of the possessor (regardless of whether he or she was stronger or weaker) to defend the possession. As our human ancestors became more intelligent, respect for the possessor became respect for possession of those objects that had special value to the possessor.

Finally, in closing this book, one more word about the sense of justice and its role in dealing with the basic conflicts between individual and group interests. There is ample evidence to suggest that biological predispositions initiate the emotions involved in judgments of fair and unfair. Emotions can turn into passion and often reason alone does not suffice to constrain them. This is where the ideal of justice can become one of the most powerful factors in law enforcement. At times it can be the glue that holds the legal system together. People need to believe in justice in order to follow the law. Yet a person's individual sense of justice, this brain mechanism that I discussed in a previous chapter, might give signals about what is fair in a given situation that differ from the concept of justice adopted by the society in which one lives.

An ideal does not have to conform to reality. The individual sense of justice and a person's desire to live in a society believed to be just, however, is not an ideal but a biological reality rooted in our human nature.

NOTES

1. There are precursors of a social machinery in nonhuman primates. Slides shown by D. A. Hamburg, Stanford University Lecture Series, in October 1973, may illustrate this. His observations revealed that dominant males will attempt to enforce the rules of the group, even at times interfering with mother-infant relationships when the mother's behavior results in cries and whimpering by the infant. For example, in one incident observed at the Gombe Research Center, an adult female could not keep up with the progress of the troop on the daily march because her infant was too sick to cling to her. This led to a conflict between very powerful drives in the animal: proper care of the infant and staying with the troop on the march. This conflict influenced not only the conduct of the baboon mother but also that of the troop leader who attempted to herd her. When the mother tried to carry the infant in her hand, she was unable to follow the troop without hurting the infant, which was pushed to the ground with each step. The whimpers of the infant immediately caught the attention of the dominant male, who approached the mother and threatened her. The mother picked up the infant and again tried to follow, but the infant's

cries brought the same results. After several attempts the mother resolved her conflict by abandoning the dying infant.

2. Piers, M. W. (1978. *Infanticide: Past and present*. New York: Norton), gives a historical view of the underlying reasons for infanticide and the cruel punishments of unwed mothers. "Discrimination against the unwed mother runs like a red thread through the history of Western civilization."

3. Piers calls it "a tale of seduction, pregnancy, infanticide, and death penalty." Supra note 2.

4. *Peninsula Times Tribune* reprint of July 16, 1987:

> *Court refuses to order sentence for Vacaville woman in infanticide.* San Francisco. The State Supreme Court refused Wednesday to order a prison sentence for a young Vacaville woman convicted of murdering her newborn baby to hide it from her boyfriend.
>
> None of the justices voted to grant a hearing on a prosecution appeal from a lower-court ruling upholding the second-degree murder conviction of Lisa Nakatani but setting aside her prison sentence of 15 years to life.

The action returns the case to Solano County Superior Judge Dwight Ely for a probation order that could include up to a year in county jail.

Superior Court #C17867. Probation Order: 5 years, fine $300.00, counseling/treatment, psychotherapy.

5. Bücheler, Zitelmann, supra note 8, Chapter 2.

References

Adams, R. M. 1960. The origin of cities. *Scientific American* 9:153, 160, 161.

Alexander, R. 1986. Biology and law. In *Ostracism: A social and biological phenomenon*, eds. M. Gruter and R. D. Masters. New York: Elsevier Science.

Alexander, R. D. 1987. *The biology of moral systems*. New York: Aldine de Gruyter.

Aronson, E. ed. 1981. *The social animal*. San Francisco: W. H. Freeman.

Axelrod, R. 1984. *The evolution of cooperation*. New York: Basic Books.

Bachofen, J. J. (1869). *Mutterrecht und Urreligion*.

Beach, F. A. 1965. *Sex and behavior*. New York: E. Wiley.

Beauvoir, S. de. 1971. *The second sex*.

Beckstrom, J. H. 1985. *Sociobiology and the law*. Chicago: University of Illinois Press.

Bender, S. 1974. Privacies of life. (March) *Harper's*, 63.

Bodenheimer, E. 1988. Law as a bridge between is and ought. *Ratio Juris*. 1:2.

Bohannan, P. 1969. Ethnography and comparison in legal anthropology. In *Law in culture and society*, ed. L. Nader. Chicago: Aldine.

Bohannan, P. (in press). *We the alien*.

Boswell, J. [1799] 1953. *Life of Johnson*. ed. R. W. Chapman. Oxford: Oxford University Press.

Bowlby, J. 1969. *Attachment and loss, Vol. 1, Attachment*. New York: Basic Books.

Braidwood, R. J. 1960. The agricultural revolution. *Scientific American* 9:131.

Brecht, A. 1959. *Political theory*. Princeton, NJ: Princeton University Press.

Bücheler, F., and E. Zitelmann. [1885] 1974. *The law of Gortyn (Das Recht von Gortyn)*.

Buxbaum, D. 1968. *Family law and customary law in Asia*. The Hague, Netherlands: Martinus Nijhoff.

Buydens-Branchey, L., M. H. Branchey, D. Noumair, et al. Age of alcoholic onset. II. Relationship of susceptibility to serotonin precursor availability. *Arch. Gen. Psychiat.* 46:231-236

Calabresi, G., and P. Bobbitt. 1978. *Tragic choices*. New York: W. W. Norton.

Calabresi, G. 1985. *Ideals, beliefs, attitudes and the law*. Syracuse, NY: Syracuse University Press.

Capron, A. M. 1987. Alternative birth technologies: Legal challenges. *University of California—Davis Law Review*, 20:679.

Cashman, S. D. 1981. *Prohibition: The lie of the land*. New York: Free Press.

Coase, R. H. 1960. The problem of social cost. *J. of Law & Economics* 3:1

Cooter, R. 1990. *Inventing property: Economic theories of the origins of market property applied to Papua New Guinea*. Olin Foundation Working Paper #88-5, University of Virginia.

Cooter, R., and T. Ulen. 1988. *Law and economics*. Glenview, IL: Scott, Foresman.

Darwin, C. [1859] 1964. *On the origin of species*. New York: Atheneum.

Dawkins, P. 1986. *The blind watchmaker*. New York: W. W. Norton.

DeVore, I. 1972. *Quest for the roots of society: The marvels of animal behavior*. Washington DC: National Geographic Society.

Djerassi, C. 1987. The politics of contraception—The view from Tokyo. *Technol. Soc.* 9:157.

Djerassi, C. 1989. The bitter pill. *Science* 245:28.

Dorfman, R., and N. S. Dorfman, eds. 1977. *Economics of the environment*. New York: W. W. Norton.

Dubos, R. 1980. *Man adapting*. New Haven, CT: Yale University Press.

Dworkin, R. 1977. *Taking rights seriously*. Cambridge, MA: Harvard University Press.

Edey, M. 1972. *The missing link—The emergence of man*. New York: Time-Life Books.

Ehrlich, E. [1913] 1975. *Grundlegung der Soziologie des Rechts*. Translated as *Fundamental principals of the sociology of law*. New York: Arno.

Ehrlich, P. 1990. *Population explosion*. New York: Simon & Schuster

Eibl-Eibesfeldt, I. 1972. *Die !Ko-Buschmann-Gesellschaft*. Munich, Germany: R. Piper.

Eibl-Eibesfeldt, I. 1984. *Die Biologie des menschlichen Verhaltens*. Munich: R. Piper.

Eisenberg, J. F. 1972. The elephant: Life at the top. In *The marvels of animal behavior*.

Ellickson, R. C. 1986. Of Coase and cattle: Dispute resolution among neighbors in Shasta County. *Stanford Law Review* 38:623.

Elliott, E. D. 1985. The evolutionary tradition in jurisprudence. *Columbia Law Review*, 85:1

Else, J. G., and P. C. Lee, eds. 1986. *Primate ontogeny, cognition and social behavior*. Cambridge, UK: Cambridge University Press.

Epstein, R. 1980. A taste for privacy? Evolution and the emergence of a naturalistic ethic. *Journal of Legal Studies* 9:665.

Espinas, A. 1878. *Über die tierischen Societäten*. Paris, France: Bailliere.

Essock-Vitale, S., and M. T. McGuire. 1980. Predictions derived from the theories of kin selection and reciprocation assessed by onthropological data. *Ethology and Sociobiology* 1:233-243.

Feierman, J. R. ed. 1990. *Pedophilia*. New York: Springer Verlag.

Finkelhor, D. 1986. *A sourcebook on child sexual abuse. Beverly Hills: Sage.*

Flint, J., et al. eds. 1990. *Tearing down the curtain*. London: Hodder & Stoughton.

Fox, R. 1971. *Kinship and marriage*. Baltimore, MD: Penguin Books.

Fox, R. 1983. *Red lamp of incest*. South Bend, IN: University of Notre Dame Press.
Fox, R. N. d. *In the matter of plural marriage, Reynolds v. USA 1879 and Cleveland vs. USA 1946*.
Friedan, B. 1974. *The feminine mystique*. New York: W. W. Norton.
Friedman, L. 1969. Legal culture and social developments. *Law and Society Review* 4:19.
Friedrich, C. J. 1969. *The philosophy of law in historical perspective*. Chicago: University of Chicago Press.
Fuchs, L. H. 1972. *Family matters*. New York: Random House.
Galante, M. A. 1986. Courts not wed to palimony. Who can recover? *National Law Journal* 45 col in v8 (July 14):3 col. 3.
Gazzaniga, M. S. 1985. *The social brain*. New York: Basic Books.
Gazzaniga, M. S. 1988. *Mind matters*. Boston, MA: Houghton Mifflin.
Gilligan, C. 1982. *In a different voice*. Cambridge, MA: Harvard University Press.
Ginsburg, B., and F. Carter, eds. 1987. *Premenstrual syndrome*. New York: Plenum.
Glendon, M. A. 1987. *Abortion and divorce in Western law*. Cambridge, MA: Harvard University Press.
Goodall, J. 1986. Social rejection, exclusion, and shunning among the Gombe chimpanzees. In *Ostracism: A social and biological phenomenon*, eds. M. Gruter and R. D. Masters. New York: Elsevier Science.
Goodall, J. van Lawick-. 1971. *In the shadow of man*. Boston, MA: Houghton Mifflin.
Goodall, J. van Lawick-. 1968. The behavior of free living chimpanzees in the Gombe Stream Reserve. *Animal Behavior Monographs* 1:161.
Gruter, M. 1944. *Die Stellung der Ehefrau im englischen Scheidungsrecht*. Doctor of Jurisprudence dissertation, University of Heidelberg, Heidelberg, Germany.
Gruter, M. 1976. Die Bedeutung der Verhaltensforschung für die Rechtswissenschaft. In *Schriftenreihe zur Rechtssoziologie und Rechtstatsachenforschung*, Bd. 36, eds. E. E. Hirsch and M. Rehbinder. Berlin: Duncker & Humblot.
Gruter, M. 1977. Law in sociobiological perspective. *Florida State University Law Review* 5:2.
Gruter, M. 1982. Biologically based behavioral research and the facts of law. *J. Soc. Biol. Struct*. 5:315-323.
Gruter, M., and P. Bohannan, eds. 1983. *Law, biology and culture: The evolution of law*. Santa Barbara, CA: Ross Erikson.
Gruter, M., and R. D. Masters, eds. 1986. *Ostracism: A social and biological phenomenon*. New York: Elsevier Science.
Gruter, M. 1986. Ostracism on trial. In *Ostracism: A social and biological phenomenon*, eds. M. Gruter and R. D. Masters. New York: Elsevier Science.
Gruter Institute. 1988. In the matter of Baby M, *Amicus Curiae Brief*.
Hahn, R. W., and G. L. Hester. 1989. Where did all the markets go? An analysis of EPA's emissions trading program. 6 *Yale J. Reg*. 109.
Hall, E. T. 1977. *Beyond culture*. New York: Doubleday.
Hamilton, W. D. 1964. The genetic evolution of social behavior. *J. Theor. Biol.* 7:1-16.
Hamilton, W. D. 1964. *J. Theor. Biol*. 7:1-64.
Hardin, G. J. 1968. The tragedy of the commons. *Science* 162:1243

Harlow, H. 1959. Love in infant monkeys. *Scientific American* 6:68.

Hassenstein, B. 1973. *Verhaltensbiologie des Kindes*. Munich: R. Piper.

Hirshleifer, J. 1980. Privacy: Its origin, function and future. *J. Legal Stud.* 9:649.

Hirth, K. 1984. Xochicalco: Urban growth and state formation in Ccentral Mexico. *Science* 255:579-586.

Hoebel, B. G. 1983. The neural and chemical basis of reward. In *Law, biology and culture,* eds. M. Gruter and P. Bohannan. Santa Barbara, CA: Ross Erikson.

Hoebel, E. A. 1954. *Law of primitive man*. Cambridge: Harvard University Press.

Hofstader, R. 1955. *Social Darwinism in American thought* rev. ed. Boston: Beacon Press.

Howard, G. E. II, 1904. *A history of matrimonial institutions.* Chicago: University of Chicago Press.

Hurlbut, C. 1968. *Minerals and man*. New York: Random House.

Johanson, D., and M. Edey. 1981. *Lucy*. New York: Simon & Schuster.

Johnson, R. 1972. *Aggression in man and animals*. Philadelphia: W. B. Saunders.

Jolly, A. 1972. *The evolution of primate behavior*. New York: Macmillan.

Krebs, J. R., and N. B. Davies. 1987. *An introduction to behavioral ecology*. Sunderland, MA: Sinauer Associates.

Kummer, H. 1971. *Primate societies*. New York: Aldine Atherton.

Kyrig, D. E., ed., 1985. *Law, alcohol and order: Perspectives on national prohibition.* Westport, CT: Greenwood.

Lancaster, J. B., J. Altmann, A. S. Rossi, L. R. Sherrod, eds., 1987. *Parenting across the life span: Biosocial dimensions*. New York: Aldine de Gruyter.

Leakey, L., and V. Goodall. 1969. *Unveiling man's origins*. Cambridge, MA: Schenkman.

Lévi-Strauss, C. 1969. *The elementary structures of kinship*. Boston, MA: Beacon.

Lewis, D. B., and D. M. Gower. 1980. *Biology of communication*. New York: John Wiley.

Lieberman, P. 1984. *The biology and evolution of language*. Cambridge, MA: Harvard University Press.

Lorenz, K. 1952. *King Solomon's ring*. New York: Y. Crowell.

Lorenz, K. 1963. *Das sogenannte Böse*. Wien, Austria: Dr. G. Borotha-Schoeler.

Lorenz, K. 1965. *Evolution and modification of behavior*. Chicago: University of Chicago Press.

Lorenz, K. 1966. *On aggression*. New York: Bantam Books.

MacDonald, K. B. 1988. *Social and personality development*. New York: Plenum Press.

MacDonald, K. B. ed. 1988. *Sociobiological perspectives on human development*. New York: Springer Verlag.

MacLean, P. D. 1983. A triangular brief on the evolution of brain and law. In *Law, biology and culture,* eds. M. Gruter and P. Bohannan. Santa Barbara, CA: Ross Erikson.

Maine, H. 1861. *Ancient law*. London, UK: J. Murray.

Malinowski, B. [1926] 1972. *Crime and custom in savage society*. Totowa, NJ: Littlefield, Adams.

Malinowski, B. 1929. *The sexual life of savages*. New York: Harcourt, Brace & World.

Manheim, R. 1967. (trans.). *Myth, religion, and mother right: Selected writings of J. J. Bachofen*. Bollingen Series No. 84. Princeton University Press.

Marler, P. R. 1972. The drive to survive. In *The marvels of animal behavior.*

Massell, G. 1968. Law as an instrument of revolutionary change in a traditional milieu: The case of Soviet Central Asia. *Law & Society Review* II:2.

Masters, R. D. 1977. Nature, human nature, and political thought. In *Human nature in politics,* eds. R. Pennock and J. Chapman. New York: New York University Press,

Masters, R. D. 1989. *The nature of politics.* New Haven, CT: Yale University Press.

Masters, W., and V. Johnson. 1970. *Human sexual inadequacy.* Boston, MA: Little, Brown.

McGuire, M. T., and M. Raleigh. 1988. *Legal concepts in the face of research on the neurochemical substrates of social behavior,* unpublished manuscript.

McGuire, M. T., M. J. Raleigh, and C. Johnson. 1983. Social dominance in adult vervet monkeys I: General considerations. *Soc. Sci. Information* 22:89-121.

Meijer, N. J. 1971. *Marriage law and policy in the Chinese People's Republic.* Hong Kong: Hong Kong University Press.

Miller Weisberger, J. 1986. Marital property discrimination: Reform for legally excluded women. In *Ostracism: A social and biological phenomenon,* eds. M. Gruter and R. D. Masters. New York: Elsevier Science.

National Academy of Sciences, Board of Agriculture. 1986. *Pesticide resistance: Strategies and tactics for management.*

Parsons, T. 1964. Introduction. In Weber, *The theory of social and economic organization.*

Piaget, J. 1959. *The language and thought of the child.* London, UK: Routledge and Kegan Paul.

Piers, M. W. 1978. *Infanticide: Past and present.* New York: W. W. Norton.

Plomin, R., 1990. The role of inheritence in behavior. *Science* 248:117-122.

Podgorecki, A. 1981. Living law. *Law & Society Rev.* 15:1.

Polinski, A. M. 1989. *An introduction to law and economics.* Boston, MA: Little, Brown.

Potts, R. 1988. *Early hominid activities at Olduvai.* New York: Aldine de Gruyter.

Raleigh, M., and M. T. McGuire. 1980. Biosocial pharmacology. *Journal McLean Hospital* 2:73-84.

Rawls, J. 1971. *A theory of justice.* Cambridge, MA: The Belknap Press of Harvard University Press.

Reader, J. 1988. *Man on Earth.* Austin: University of Texas.

Rehbinder, M. 1989. *Rechtssoziologie.* Berlin: Walter de Gruyter.

Reidinger, P. 1986. Family law: Cohabitation confers property rights. *ABA Journal* 72 (Oct l):92(2).

Reuben, R. 1988. Justices to weigh custodial rights of unwed parents. *Los Angeles Daily Journal* 30 col in v 101 (Nov. 28): pl col 4.

Rheinstein, M., ed. 1954. *Max Weber on law in economy and society.* Cambridge, MA: Harvard University Press.

Rheinstein, M. 1972. *Marriage stability, divorce and the law.* Chicago: University of Chicago Press.

Rodgers, W. 1982. Bringing people back: Toward a comprehensive theory of taking in natural resources law. *Ecology L. Q.* 10:205.

Rodgers, W. H., Jr. 1986. The evolution of cooperation in natural resources law: The drifter/habitue distinction. *University of Florida Law Review* 38 (Spring):2.

Roy, A., and M. Linnoila. 1988. Suicidal behavior, impulsiveness and serotonin. *Acta Psychiat. Scand.* 78:529-535.

Seymour-Smith, C., ed. 1986. *Dictionary of anthropology*. Boston: G. K. Hall.

Slater, P. J. B. 1985. *An introduction to ethology*. Cambridge, UK: Cambridge University Press.

Smuts, B. B., et. al., eds. 1987. *Primate societies*. Chicago: University of Chicago Press.

Sorenson, R. 1973. *Adolescent sexuality in contemporary America*. New York: World Publishing.

Strauss, L. 1953. *Natural right and history*. Chicago: University of Chicago Press.

Sumner, W. G. 1963. In *Social Darwinism: Selected essays of W. G. Sumner*, ed. S. Pearsons. Englewood Cliffs, NJ: Prentice Hall.

Taylor, C. E., and M. T. McGuire. 1988. Reciprocal altruism: 15 years later. *Ethology and Sociobiology* 9:2-4

Teleki, G. 1973. The omnivorous chimpanzee. *Scientific American* 33 (Jan.):42.

Tietze, C., and S. K. Henshaw. 1986. *Induced abortion: A world review*. New York: Guttmacher.

Tiger, L. 1988. *The manufacture of evil*. New York: Harper & Row.

Tiger, L., and R. Fox. [1971] 1989. *The imperial animal*. New York: H. Holt.

Tinbergen, N. 1974. Ethologie. In *Kritik der Verhaltensforschung: Konrad Lorenz und seine Schule*, ed. G. Roth. Munich: C. H. Beck.

Trivers, R. 1971. The evolution of reciprocal altruism. *Quarterly Review of Biology* 46.

Trussell, J. 1988. Teenage pregnancies in the United States. *Fam. Plan. Perspect.* 20:262.

Waal, F. de 1986. The brutal elimination of a rival among captive male chimpanzees. In *Ostracism: A social and biological phenomenon*, eds. M. Gruter and R. D. Masters. New York: Elsevier Science.

Waddington, C. H. 1967. *The ethical animal*. Chicago: University of Chicago Press.

Wahl, E. 1973. Influences Climatiques sur l'Evolution du Droit en Orient en Occident. *Revue International de Droit Comparé*.

Washburn, S. C., and D. Hamburg. 1972. Aggressive behavior in Old World monkeys and apes. In *Primate patterns*, ed. P. Dolhinow. New York: Holt, Rinehart and Winston.

Wickler, W. 1972. *The sexual code*. Garden City, NY: Doubleday.

Wills, W. H. 1988. *Early prehistoric agriculture in the American southwest*. Santa Fe, NM: School of American Research Press.

Wilson, E. O. 1975. *Sociobiology*. Cambridge, MA: Harvard University Press.

Winick, C., ed. 1970. *Dictionary of anthropology*. Totowa, NJ: Littlefield, Adams.

Wittfogel, K. 1957. *Oriental despotism*. New Haven, CT: Yale University Press.

Ziegenfuss, W. 1956. *Handbuch der Soziologie*. Stuttgart, Germany: Ferdinand Enke Verlag.

About the Author

Margaret Gruter, born in Germany, was educated in law and holds a Doctor of Jurisprudence degree from the University of Heidelberg. In 1951 she immigrated to the United States, where she and her physician husband managed a medical facility in rural Ohio for a dozen years. Returning to her academic interests, she took courses in the fields of biology and law and received an advanced degree (J.S.M.) from Stanford University Law School. She taught at Stanford and Heidelberg Universities and lectured at several German and Swiss universities. Her books and articles have been published in German as well as in English.

In the early 1980s, she founded the *Gruter Institute for Law and Behavioral Research*, which examines relationships between law and biology, with an emphasis on how recent findings in biology and anthropology inform the history, practice, understanding, and utility of law. She continues her research on the interaction between law and biologically based human behavior, together with associates of the Institute. The Gruter Institute is the only one of its type in the world, and it continues to remain at the intellectual forefront of biology-law explorations.

NOTES